# Quantitative Perspectives on Behavioral Economics and Finance

**Series Editor**
James Ming Chen
College of Law
Michigan State University
East Lansing, MI, USA

The economic enterprise has firmly established itself as one of evaluating human responses to scarcity not as a rigidly rational game of optimization, but as a holistic behavioral phenomenon. The full spectrum of social sciences that inform economics, ranging from game theory to evolutionary psychology, has revealed the extent to which economic decisions and their consequences hinge on psychological, social, cognitive, and emotional factors beyond the reach of classical and neoclassical approaches to economics. Bounded rational decisions generate prices, returns, and resource allocation decisions that no purely rational approach to optimization would predict, let alone prescribe. Behavioral considerations hold the key to long-standing problems in economics and finance. Market imperfections such as bubbles and crashes, herd behavior, and the equity premium puzzle represent merely a few of the phenomena whose principal causes arise from the comprehensible mysteries of human perception and behavior. Within the heterodox, broad-ranging fields of behavioral economics, a distinct branch of behavioral finance has arisen. Finance has established itself as a distinct branch of economics by applying the full arsenal of mathematical learning on questions of risk management. Mathematical finance has become so specialized that its practitioners often divide themselves into distinct subfields. Whereas the $P$ branch of mathematical finance seeks to model the future by managing portfolios through multivariate statistics, the $Q$ world attempts to extrapolate the present and guide risk-neutral management through the use of partial differential equations to compute the proper price of derivatives. The emerging field of behavioral finance, worthy of designation by the Greek letter psi ($\psi$), has identified deep psychological limitations on the claims of the more traditional $P$ and $Q$ branches of mathematical finance. From Markowitz's original exercises in mean-variance optimization to the Black-Scholes pricing model, the foundations of mathematical finance rest on a seductively beautiful Gaussian edifice of symmetrical models and crisp quantitative modeling. When these models fail, the results are often catastrophic. The $\psi$ branch of behavioral finance, along with other "postmodern" critiques of traditional financial wisdom, can guide theorists and practitioners alike toward a more complete understanding of the behavior of capital markets. It will no longer suffice to extrapolate prices and forecast market trends without validating these techniques according to the full range of economic theories and empirical data. Superior modeling and data-gathering have made it not only possible, but also imperative to harmonize mathematical finance with other branches of economics. Likewise, if behavioral finance wishes to fulfill its promise of transcending mere critique and providing a more comprehensive account of financial markets, behavioralists must engage the full mathematical apparatus known in all other branches of finance. In a world that simultaneously lauds Eugene Fama's efficiency hypotheses and heeds Robert Shiller's warnings against irrational exuberance, progress lies in Lars Peter Hansen's commitment to quantitative rigor. Theory and empiricism, one and indivisible, now and forever.

More information about this series at
http://www.palgrave.com/gp/series/14524

Nicholas L. Georgakopoulos

# Illustrating Finance Policy with *Mathematica*

Nicholas L. Georgakopoulos
Indiana University
Indianapolis, IN, USA

Quantitative Perspectives on Behavioral Economics and Finance
ISBN 978-3-319-95371-7        ISBN 978-3-319-95372-4    (eBook)
https://doi.org/10.1007/978-3-319-95372-4

Library of Congress Control Number: 2018949044

Cover image: © winivic, iStock/Getty Images
Cover design by Tjasa Krivec

This Palgrave Macmillan imprint is published by the registered company Springer Nature Switzerland AG
The registered company address is: Gewerbestrasse 11, 6330 Cham, Switzerland

*To mental health professionals,
whose value is underscored
by the apparent utter
rationality of finance.*

# Foreword

*Seeing Is Believing*

Truth routinely manifests itself through mathematics. "[T]he real world may be understood in terms of the real numbers, time and space and flesh and blood and dense primitive throbbings sustained somehow and brought to life by a network of secret mathematical nerves ...."[1] Just as law uses words to animate the "enterprise of subjecting human conduct to the governance of rules,"[2] "[n]ature talks to us in the language of mathematics."[3] Through this volume, *Illustrating Finance Policy with* Mathematica, Nicholas L. Georgakopoulos enriches the series, *Quantitative Perspectives on Behavioral Economics and Finance*, by demonstrating the mathematical underpinnings of law and finance in vivid, visual terms.

---

[1] DAVID BERLINSKI, A TOUR OF THE CALCULUS, at xiii (1996).

[2] LON L. FULLER, THE MORALITY OF LAW 122 (rev. ed. 1969).

[3] Peter Hilton, *The Mathematical Component of a Good Education, in* MISCELLANEA MATHEMATICA 145–554, 149 (Peter Hilton, Friedrich Hirzebruch & Reinhold Remmert eds., 1991); *accord* Peter Hilton, *Foreword: Mathematics in Our Culture, in* JAN GULLBERG, MATHEMATICS: FROM THE BIRTH OF NUMBERS, at xvii–xxii, xix (1997).

Professor Georgakopoulos seeks to provide "a visual explanation" of concepts such as "the CAPM [capital asset pricing model] or the call options pricing formula" in intuitive and memorable form. Indeed, he acknowledges that this book, at its "most basic," enables the reader "to learn the fundamental concepts of modern finance without the quantitative foundation that most finance books require," by "getting the intuitions visually from the graphics." To do so, however, neglects this book's full quantitative and computational potential. This preface therefore encourages readers to aspire, however slowly, toward Prof. Georgakopoulos's "more demanding approach" of learning "how to use Mathematica in simple finance applications and in the production of graphics."

Both Prof. Geogakopoulous and I hold academic appointments as professors of law. This volume treats lawyers, law students, and their instructors as members of its target audience. Admittedly, "[i]t is an open secret that lawyers" (stereo)typically "don't like math."[4] In a legal culture whose leaders shamelessly confess their ignorance of the "fine details of molecular biology,"[5] lawyers and lawmakers run a dire risk of falling behind "the extraordinary rate of scientific and other technological advances that figure increasingly in litigation" and, for that matter, in daily life.[6]

Even though law addresses subjects "so vast that fully to comprehend [them] would require an almost universal knowledge ranging from geology, biology, chemistry and medicine to the niceties of the legislative, judicial and administrative processes of government,"[7] the "extraordinary condition" of the legal profession "makes it possible for [someone] without any knowledge of even the rudiments of chemistry to pass

---

[4] Lisa Milot, *Illuminating Innumeracy*, 63 Case W. L. Rev. 769–812, 769 (2013).

[5] Association for Molecular Pathology v. Myriad Genetics, Inc., 133 S. Ct. 2107, 2120 (2013) (Scalia, J., concurring in part and concurring in the judgment) ("I join the judgment of the Court, and all of its opinion except Part I-A and some portions of the rest of the opinion going into fine details of molecular biology. I am unable to affirm those details on my own knowledge or even my own belief").

[6] Jackson v. Pollion, 753 F.3d 786, 788 (7th Cir. 2013) (Posner, J.).

[7] Queensboro Farms Prods., Inc. v. Wickard, 137 F.2d 969, 975 (2nd Cir. 1943) (describing agriculture, particularly dairy farming).

upon" scientifically or technologically sophisticated questions.[8] No less than other social sciences, law should aspire to a level of "numeracy," one that is "less about numbers per se and more about statistical inference or how to interpret and understand scientific ... studies."[9]

A mastery of basic mathematical concepts should serve as a foundation for serious legal scholarship. As a group that not only digests but also delivers postmodern criticism,[10] legal scholars can surely grasp mathematics, which after all is merely another branch of philosophy.[11] "Legal reasoning," in particular, represents merely a special case of "*theory construction*."[12] The prospect of teaching machines to speak a language that expresses and conveys legal knowledge fulfills the aesthetic if not the practical goals of information theory.[13]

Despite the law's reputation for quantitative ineptitude, mathematical thinking naturally suits this discipline. Social scientists have nurtured "something like a third culture" between science and literature in order to improve the circumstances under which real "human beings are living."[14] Accordingly, scholars within this tradition enjoy a special

---

[8] Parke-Davis & Co. v. H.K. Mulford Co., 189 F. 95, 115 (S.D.N.Y. 1911) (Hand, J.).

[9] Edward K. Cheng, *Fighting Legal Innumeracy*, 17 GREEN BAG 2D 271–78, 272 (2014); *accord* United States *ex rel.* Customs Fraud Investigations, LLC. v. Victaulic Co., 839 F.3d 242, 270 & n.56 (3rd Cir. 2016), cert. denied, 138 S. Ct. 107 (2017). *See generally* JOHN ALLEN PAULOS, INNUMERACY: MATHEMATICAL ILLITERACY and ITS CONSEQUENCES (2nd ed. 2001) (1st ed. 1988).

[10] *Cf.* STANLEY EUGENE FISH, DOING WHAT COMES NATURALLY: CHANGE, RHETORIC, AND THE PRACTICE OF THEORY IN LITERARY AND LEGAL STUDIES (1990).

[11] *See* U.S. PATENT & TRADEMARK OFFICE, GENERAL REQUIREMENTS BULLETIN FOR ADMISSION TO THE EXAMINATION FOR REGISTRATION TO PRACTICE IN CASES BEFORE THE UNITED STATES PATENT AND TRADEMARK OFFICE 37 (2008) (describing mathematics as a philosophical discipline and therefore insufficient by itself to satisfy the technical training requirement for eligibility to take the Patent and Trademark Office examination); *see also* 37 C.F.R. § 11.7(a)(2)(ii) (requiring practitioners before the USPTO to "[p]ossess the legal, scientific, and technical qualifications necessary ... to render ... valuable service" to patent and trademark applicants).

[12] L. Thorne McCarty, *Some Arguments About Legal Arguments*, *in* SIXTH INTERNATIONAL CONFERENCE ON ARTIFICIAL INTELLIGENCE AND LAW 215–24, 221 (1997) (emphasis in original).

[13] *See generally* ABRAHAM MOLES, INFORMATION THEORY AND ESTHETIC PERCEPTION (Joel E. Cohen trans., 1968); FRIEDER NAKE, ÄSTHETIK ALS INFORMATIONSVERARBEITUNG (1974) ("Aesthetics as Information Processing").

[14] C.P. SNOW, THE TWO CULTURES: AND A SECOND LOOK 70 (2nd ed. 1965).

opportunity to unite the literary culture's "canon of works and expressive techniques" with the scientific culture's "guiding principles of quantitative thought and strict logic."[15] At their best, social scientists bridge all of contemporary civilization's intellectual subcultures.[16]

The application of empirical methods to discrete problems is perhaps the most familiar and deeply rooted form of mathematically informed social science. More than a century after Oliver Wendell Holmes declared that "the man of the future is the man of statistics and the master of economics,"[17] and three decades after Richard Posner celebrated the decline of law as an autonomous discipline,[18] empiricism dominates contemporary legal studies.

Law and finance face staggering amounts of data and fierce competition among analytical models. Imperfectly articulated hypotheses in these branches of social science will likely lead neither to elegant closed-form solutions nor to pathological functions. Even the most thoughtfully elaborated empirical tests fail to deliver complete knowledge about law and its underlying logic. "Every year, if not every day, we have to wager our salvation upon some prophecy based upon imperfect knowledge."[19]

Mathematical analysis in social science typically follows a dialectic of romance, frustration, and eventual reconciliation of internal logic with

---

[15] Frank Wilczek, *The Third Culture*, 424 NATURE 997–98, 997 (2003).

[16] *Cf. Cultural Divides, Forty Years On*, 398 NATURE 91, 91 (1999) (recognizing how C.P. Snow's depiction of two cultures "still resonates" in a world "where cultural antipathies are very much alive and kicking").

[17] Oliver Wendell Holmes, *The Path of the Law*, 10 HARV. L. REV. 457–78, 470 (1897), *reprinted in* 110 HARV. L. REV. 991–1009, 1001 (1997).

[18] *See generally* Richard A. Posner, *The Decline of Law as an Autonomous Discipline: 1962–1987*, 100 HARV. L. REV. 761–80 (1987).

[19] Abrams v. United States, 250 U.S. 616, 630 (1919) (Holmes, J., dissenting).

external reality. All of science follows a familiar progression. "Normal science does not aim at novelties of fact or theory and, when successful, finds none."[20] But "fundamental novelties of fact and theory" trigger "the recognition that nature has somehow violated the paradigm-induced expectations that govern normal science."[21] Once an "awareness of anomaly ha[s] lasted so long and penetrated so deep" as to plunge a scientific discipline into "a state of growing crisis," a succeeding "period of pronounced professional insecurity" over "the persistent failure of the puzzles of normal science" prompts a fruitful search for new rules.[22] Wonderfully, "random shocks"—the impetus for progress in economics, law, and science—happen to be "the subject matter" of finance and its regulation.[23]

Mathematics serves as a source of beauty and sensory delight. Uniquely among human endeavors, mathematics boasts "a beauty cold and austere, ... without any appeal to any part of our weaker nature, without the gorgeous trappings of painting or music, yet sublimely pure, and capable of a stern perfection such as only the greatest art can show. The true spirit of delight, the exaltation, the sense of being more than Man, which is the touchstone of the highest excellence, is to be found in mathematics as surely as poetry."[24] As the poet Edna St. Vincent Millay expressed the sentiment: "Euclid alone has looked on Beauty bare."[25] What Justice Potter Stewart said of obscenity (that he knew it when he saw it)[26] finds a parallel in Paul Erdős's definition of

---

[20] *See* THOMAS S. KUHN, THE STRUCTURE OF SCIENTIFIC REVOLUTIONS 52 (2d ed. enlarged, 1970).

[21] *Id.* at 52–53.

[22] *See id.* at 66–67.

[23] John Y. Campbell, *Asset Pricing at the Millennium*, 55 J. FIN. 1515–67, 1515 (2000).

[24] BERTRAND RUSSELL, *The Study of Mathematics, in* MYSTICISM AND LOGIC, AND OTHER ESSAYS 58–73, 60 (1988); *accord* Jim Chen, *Truth and Beauty: A Legal Translation*, 41 U. TOLEDO L. REV. 261–67, 265 (2010).

[25] EDNA ST. VINCENT MILLAY, *Euclid Alone Has Looked on Beauty Bare, in* SELECTED POEMS 52 (J.D. McClatchy ed., 2003).

[26] *See* Jacobellis v. Ohio, 378 U.S. 184, 197 (1964) (Stewart, J., concurring) ("[P]erhaps I could never succeed in intelligibly [defining obscenity]. But I know it when I see it ....").

mathematical beauty: "Why are numbers beautiful? It's like asking why Beethoven's Ninth Symphony is beautiful. If you don't see why, someone can't tell you."[27]

Seeing is believing. So is hearing. Standard mathematical specifications of human sensory perception, particularly sight and sound, literally round out our world. For example, the HSV and HSL models project color onto cylindrical or conic space where *hue* (distinctive shades spanning red, magenta, blue, cyan, green, and yellow) is represented as a circular spectrum; *saturation* (the richness of color, or its chroma) is represented as linear or angular distance from a neutral, gray axis; and *value* indicates the degree of pure brightness or its opposite, darkness, relative to the origin.[28]

Music can be represented in a similar physical space. If intensity of a tone is indicated along a line segment of length 1 from the threshold of hearing (0 dB) to the threshold of pain (130 dB),[29] a corresponding angle of $\pi/2$ radians within a "unit cone" can indicate frequency across the range of human hearing, 20–20,000 Hz.[30] Twelve equally spaced notes within each pitch class (a single octave, separated by the one above it by a doubling of frequency and the one below it by a halving of frequency) appear at an interval of 100 cents within the appropriately named *chromatic scale*.[31] As sight or sound, color and music are mathematics made flesh.[32]

---

[27] PAUL HOFFMAN, THE MAN WHO LOVED ONLY NUMBERS: THE STORY OF PAUL ERDŐS AND THE SEARCH FOR MATHEMATICAL TRUTH 42 (1998) (quoting Erdős); *accord* KEITH DEVLIN, THE MATH GENE: HOW MATHEMATICAL THINKING EVOLVED AND WHY NUMBERS ARE LIKE GOSSIP 140 (2000).

[28] *See* Alvy Ray Smith, Color gamut transform pairs, 12(3) COMPUTER GRAPHICS 12–19 (August 1978) (describing the "hexcone" model of HSV colorspace); George H. Joblove & Donald Greenberg, *Color Spaces for Computer Graphics*, 12(3) COMPUTER GRAPHICS 20–25 (August 1978) (describing the HSL model and comparing it to HSV); *cf.* Dorothy Nickerson, *History of the Munsell Color System*, 1 COLOR RESEARCH & APPLIC. 121–30 (1976). *See generally* STEVEN K. SHEVELL, THE SCIENCE OF COLOR 202–06 (2d ed. 2003).

[29] *See, e.g.,* DAVID HOWARD & JAMIE ANGUS, ACOUSTICS AND PSYCHOACOUSTICS § 2.3, at 80–82 (2012); DEBI PRASAD TRIPATHY, NOISE POLLUTION 35 (2008).

[30] *See generally* HARRY F. OLSON, MUSIC, PHYSICS AND ENGINEERING 248–51 (1967).

[31] *See* 1 BRUCE BENWARD & MARILYN SAKER, MUSIC: IN THEORY AND PRACTICE 47 (7th ed. 2003).

[32] *Cf.* John 1:14 ("And the Word was made flesh, and dwelt among us….").

The most beautiful mathematical results exhibit "a very high degree of unexpectedness, combined with inevitability and economy."[33] Deep beauty subsists in connections that first appear unrelated,[34] but upon further examination reveal "metaphoric combination[s]" that "leap[] beyond systematic placement" and "explore[] connections that before were unsuspected."[35] The most "useful and fertile combinations" of ideas "present themselves to the mind in a sort of sudden illumination, after an unconscious working somewhat prolonged," and ultimately "seem the result of a first impression."[36]

The real world, however, often inconveniently fails to align itself with mathematically beautiful models. In the face of anomalous results, even the most rigorous, comprehensively elaborated approach "can no longer understand [itself] because the theories ... of [a] former age no longer work and the theories of the new age are not yet known."[37] That challenge leaves exactly one path forward: to "start afresh as if [we] were newly come into a new world."[38]

Financial economics has undergone a crisis of precisely this sort. Much of contemporary mathematical finance, from the Capital Asset Pricing Model (CAPM) to the Black–Scholes model of option pricing,[39] Merton's distance-to-default model of credit risk,[40] the original

---

[33] G.H. HARDY, A MATHEMATICIAN'S APOLOGY 29 (1940).

[34] *See* GIAN-CARLO ROTA, THE PHENOMENOLOGY OF MATHEMATICAL BEAUTY 173 (1997).

[35] JEROME S. BRUNER, *The Conditions of Creativity*, *in* ON KNOWING: ESSAYS FOR THE LEFT HAND 17–30, 20 (1963).

[36] HENRI POINCARÉ, *Mathematical Creation*, *in* THE FOUNDATIONS OF SCIENCE: SCIENCE AND HYPOTHESIS, THE VALUE OF SCIENCE, SCIENCE AND METHOD 383–94, 391 (George Bruce Halstead trans., 1913).

[37] WALKER PERCY, *The Delta Factor*, *in* THE MESSAGE IN THE BOTTLE: HOW QUEER MAN IS, HOW QUEER LANGUAGE IS, AND WHAT ONE HAS TO DO WITH THE OTHER 3–45, 3 (1986).

[38] *Id.* at 7.

[39] *See* Fischer Black & Myron S. Scholes, *The Pricing of Options and Corporate Liabilities*, 81 J. POL. ECON. 637–54 (1973); Robert C. Merton, *The Theory of Rational Option Pricing*, 4 BELL J. ECON. 141–83 (1973).

[40] *See* Robert C. Merton, *On the Pricing of Corporate Debt: The Risk Structure of Interest Rates*, 29 J. FIN. 449 (1974).

RiskMetrics specification of value-at-risk,[41] and the Gaussian copula,[42] is built on the Gaussian "normal" distribution.[43]

These elegant models—absent elaborate modifications that ruin their spare, symmetrical form—are treacherously wrong in their reporting of the true nature of risk. Many of the predictive flaws in contemporary finance arise from reliance on the mathematically elegant but practically unrealistic construction of "beautifully Platonic models on a Gaussian base."[44] Gaussian mathematics suggests that financial returns are smooth, symmetrical, and predictable. In reality, returns are skewed[45] and exhibit heavier than normal tails.[46]

The attraction in law and finance to formal elegance reflects a love affair with the Gaussian mathematics that has traditionally dominated

[41] *See* Jorge Mina & Jerry Yi Xiao. Return to RiskMetrics: The Evolution of a Standard (2001); Jeremy Berkowitz & James O'Brien, *How Accurate Are Value-at-Risk Models at Commercial Banks?*, 57 J. Fin. 1093–111 (2002).

[42] *See* Roger B. Nelsen, An Introduction to Copulas (1999); David X. Liu, *On Default Correlation: A Copula Function Approach*, 9(4) J. Fixed Income 43–54 (March 2000).

[43] *See generally* Benoit B. Mandelbrot & Richard L. Hudson, The (Mis)Behavior of Markets: A Fractal View of Risk, Ruin, and Reward (2004).

[44] Nassim Nicholas Taleb, The Black Swan: The Impact of the Highly Improbable 279 (2007).

[45] *See, e.g.*, John Y. Campbell, Andrew W. Lo & A. Craig MacKinlay, The Econometrics of Financial Markets 17, 81, 172, 498 (1997); Felipe M. Aparicio & Javier Estrada, *Empirical Distributions of Stock Returns: European Securities Markets, 1990–95*, 7 Eur. J. Fin. 1–21 (2001); Geert Bekaert, Claude Erb, Campbell R. Harvey & Tadas Viskanta, *Distributional Characteristics of Emerging Market Returns and Asset Allocation*, 24(2) J. Portfolio Mgmt. 102–116 (Winter 1998); Pornchai Chunhachinda, Krishnan Dandepani, Shahid Hamid & Arun J. Prakash, *Portfolio Selection and Skewness: Evidence from International Stock Markets*, 21 J. Banking & Fin. 143–67 (1997); Amado Peiró, *Skewness in Financial Returns*, 23 J. Banking & Fin. 847–62 (1999).

[46] *See, e.g.*, J. Brian Gray & Dan W. French, *Empirical Comparisons of Distributional Models for Stock Index Returns*, 17 J. Bus. Fin. & Accounting 451–59 (1990); Stanley J. Kon, *Models of Stock Returns—A Comparison*, 39 J. Fin. 147–65 (1984); Harry M. Markowitz & Nilufer Usmen, *The Likelihood of Various Stock Market Return Distributions, Part 1: Principles of Inference*, 13 J. Risk & Uncertainty 207–19 (1996); Harry M. Markowitz & Nilufer Usmen, *The Likelihood of Various Stock Market Return Distributions, Part 2: Empirical Results*, 13 J. Risk & Uncertainty 221–47 (1996); Terence C. Mills, *Modelling Skewness and Kurtosis in the London Stock Exchange FT-SE Index Return Distributions*, 44 Statistician 323–34 (1995). *See generally* Terence C. Mills, The Econometric Modelling of Financial Time Series (2nd ed. 1999).

the culture of the natural and social sciences.[47] Grasping the uncomfortable truth that Gaussian models of risk and return belong to "a system of childish illusions" forces our infatuation with the seductive symmetry of traditional risk modeling to pass "like first love ... into memory."[48]

This conflict is often portrayed as an irreconcilable struggle between the romance of beauty and the realism of truth. Hermann Weyl admonished physics (and presumably all other pursuits informed by mathematics) that any necessary choice between truth and beauty should favor beauty.[49] Practical versus philosophical "conflict" over "the purpose of scientific inquiry" is "an ancient [struggle] in science."[50] Although law ordinary seeks "knowledge ... for purely practical reasons, to predict and control some part of nature for society's benefit," the knowledge unveiled through mathematical analysis "may serve more abstract ends for the contemplative soul" and "[u]ncover[] new relationships" that prove "aesthetically satisfying" insofar as they "bring[] order to a chaotic world."[51]

This tension is illusory. Mathematics itself delivers an elegant denouement. Mathematical analysis is ordinarily associated with—indeed, often equated with—the application of established empirical techniques to ever-growing bodies of data. This book, however, demonstrates that the quantitative visualization of social phenomena enjoys a far broader scope and entertains vastly deeper ambitions. In stark "contrast with soulless calculation," "[g]enuine mathematics ... constitutes one of the finest expressions of the human spirit."[52]

---

[47] *See* Nassim Nicholas Taleb, The Black Swan: The Impact of the Highly Improbable 279 (2007).

[48] Berlinski, *supra* note 1, at 239.

[49] Freeman J. Dyson, *Prof. Hermann Weyl*, 177 Nature 457–58, 458 (1956) (quoting Weyl: "My work always tried to unite the true with the beautiful, but when I had to choose one or the other, I usually chose the beautiful."); *accord* Edward O. Wilson, Biophilia 61 (1984).

[50] Sharon E. Kingsland, Modeling Nature: Episodes in the History of Population Ecology 4–5 (1985).

[51] *Id.* at 4–5.

[52] Hilton, *The Mathematical Component of a Good Education*, *supra* note 3, at 151; *accord* Hilton, *Mathematics in Our Culture*, *supra* note 3, at xxi.

The "great areas of mathematics"—including "combinatorics, probability theory, statistics," and other fields of greatest interest to social science—"have undoubtedly arisen from our experience of the world around us."[53] Law and finance apply mathematical tools "in order to systematize that experience, to give it order and coherence, and thereby to enable us to predict and perhaps control future events."[54] But scientific progress responds to "what might be called the mathematician's apprehension of the natural dynamic of mathematics itself."[55]

With unmatched power and intuitive appeal, visual demonstrations such as those provided throughout *Illustrating Finance Policy with Mathematica* show precisely how "law is not indifferent to considerations of degree."[56] The very responsiveness of quantitative measures to changing conditions—an admittedly "qualitative notion" called "robustness"[57]—is "essential for law enforcement."[58] Robustness ensures that "different judges will reach similar conclusions" when they encounter similar data.[59]

Mathematics *as doing* delivers the answers that we most pressingly seek—not simply according to the data describing the world as we find it, but also according to the own internal logic of mathematics. "[T]here is nothing in the world of mathematics that corresponds to an audience in a concert hall, where the passive listen to the active. Happily, mathematicians are all *doers*, not spectators."[60] Through its

---

[53] *Id.*

[54] *Id.*

[55] *Id.*

[56] Schechter Poultry Corp. v. United States, 295 U.S. 495, 554 (1935) (Cardozo, J., concurring).

[57] Rama Cont, Romain De Guest & Giacomo Scandolo, *Robustness and Sensitivity of Risk Measurement Procedures*, 10 QUANT. FIN. 593–606, 594 (2010).

[58] Steven Kou, Xianhua Peng & Chris C. Hyde, *External Risk Measures and Basel Accords*, 38 MATH. OPERATIONAL RESEARCH 393–417, 401 (2013).

[59] *Id.*

[60] GEORGE M. PHILLIPS, MATHEMATICS IS NOT A SPECTATOR SPORT, at vii (2005).

quest for "universal interest[s]," Prof. Georgakopoulos's introduction to finance and its visual representation may yet "catch an echo of the infinite, a glimpse of its unfathomable process, a hint of the universal law."[61]

James Ming Chen

**James Ming Chen** Justin Smith Morrill Chair in Law, Michigan State University; Of Counsel, Technology Law Group of Washington, D.C.; Visiting Scholar, Visiting Scholar, University of Zagreb, Faculty of Economics and Business (Ekonomski Fakultet, Sveučilište u Zagrebu). Special thanks to Heather Elaine Worland Chen.

---

[61] Holmes, *supra* note 17, at 478, *reprinted in* 110 Harv. L. Rev. at 1009.

# Preface

## Purpose: Visual Help

The purpose of this book is not to explain these concepts from beginning to end. Rather, I think of this book as a companion to advanced law, finance, and policy courses where students that meet new concepts need a little more of a visual explanation of those concepts. Courses discuss the Capital Asset Pricing Model (CAPM) or the call options pricing formula as mathematical concepts and equations. Most students miss the intuitive appeal of those concepts. I think that seeing the CAPM evidence in a graph (Fig. 6 in Chapter 5, p. 67) or the wedge-like solid that approximates the call option valuation formula (Fig. 4 in Chapter 6, p. 82) makes a big difference. Yet those figures appear here for the first time—I have seen them in none of the books that try to introduce these concepts.

Researchers who already understand the intuitions behind the graphs can use the book's discussion of the production of graphics. This book explains the financial usage of graphical illustrations, their intuitive appeal, and their production. The Mathematica file that produces the figures of this book is available on my personal website (nicholasgeorgakopoulos. org) under scholarship, at the entry that corresponds to this book.

## The Quantitative Finance Core

The core of this book consists of the three chapters on quantitative finance. Discounting is the object of Chapter 4. The Capital Asset Pricing Model is the object of Chapter 5. The call option valuation formula is the object of Chapter 6. Granted, these are complex matters. In a classroom setting, these topics can take far more time than the corresponding number of pages may indicate.

Whereas the mathematics are complex, they are not voluminous nor do they require mathematical manipulations. My experience is that students who enjoy math, even if they do not recall nor have had any math after high school, can handle the related concepts and formulas because the intuitions come visually from the graphs.

Instructors can elect how to use the rest of the book depending on their goals. Statistically-oriented courses would tend to opt for the chapter on illustrating statistics (Chapter 7) and probability theory (Chapter 8). Corporate courses would tend to want to include the chapter on financial statements (Chapter 9). Courses focusing on making decisions would want to include the chapter on aversion toward risk (Chapter 10). Courses that wish to focus on economic concepts would tend to include the chapter on financial crises (Chapter 11). The foundational concept of the need to justify law on a failure of the market from the perspective of the analysis of Coase (Chapter 1) is too abbreviated here for students who truly encounter it for the first time and likely would need additional supportive material (such as the two-chapter treatment it receives in my Principles and Methods of Law and Economics, 2005) if an instructor truly intends to focus on it.

## The Chapters on the Mathematica Software and Math Concepts Are Not Necessary

One can read this book with several goals. The most basic one is to learn the fundamental concepts of modern finance without the quantitative foundation that most finance books require, by getting the intuitions visually from the graphics. A more demanding approach is to learn also how to use Mathematica in simple finance applications and in the production of graphics.

In the former case, the chapters explaining Mathematica and mathematical concepts (Chapters 2 and 3) are not necessary, they can be skipped, and when, in the other chapters, the exercises call for using Mathematica, readers can try using the program of their choice. When the exercise does not call for solving an equation in symbolic form, a spreadsheet program would often suffice. The online simplified version of Mathematica at wolframalpha.com can act as a supplement, providing the capacity to solve equations in symbolic form. Other programs with the capacity for symbolic algebra, such as Maple, would also be adequate.

## Legal Citation Style

This book uses the legal style of citations. This primarily means that the volume number precedes the name of the publication, that the year comes after the full citation, and that a parenthetical explaining the relevance of the cited matter follows the citation. For example, whereas in the social sciences citation style a journal article would appear as author (year) title of article, publication name volume:page, this would appear here as author, *title of article*, volume PUBLICATION NAME page (year) (explanatory parenthetical). Notice that articles are italicized and books in small capitals.

## Acknowledgements

First, I wish to thank my research assistants in this project, Allan Griffey, Henry Robison, and John Millikan. No less important has been the help that I have received from my students who questioned and struggled with these concepts over the many years of my teaching financial concepts in my various courses. For Mathematica programming help, I must acknowledge the enormous help from numerous users of the

Mathematica group of StackExchange, mathematica.stackexchange.com. For useful comments I wish to thank George Triantaphyllou, Stephen Utz and the Oral Argument Podcast for Christian Turner's and Joe Miller's enthusiastic comments about a part of this project.

Special thanks for bringing this book to light are due to Dean and Prof. Jim Chen, the editor of this series, and Elizabeth Graber, Palgrave Macmillan's Acquisitions Editor, Assistant Editor Allison Neuburger, and Azarudeen Ahamed Sheriff, the book's production manager. Prof. Chen deserves an additional round of thanks for his colorful foreword.

Vail, Colorado, USA                                  Nicholas L. Georgakopoulos

# Summary Table of Contents

# Contents

# List of Figures

# List of Tables

# 1

# The Non-graphical Foundation: Coase and The Law's Irrelevance

The foundation of all policy analysis springs from the thought of the late Ronald Coase that individuals bargain around rules to achieve the best that they can for themselves. According to economic principles, individuals pursue the course of action that maximizes their welfare (not limited to their monetary welfare). Everybody does that in the context of a market economy, and economic forces push individual action toward greater productivity. Injecting in this system a law that impedes a preferred activity will only result in people bargaining around this law, this suboptimal allocation of rights by the legal system. As long as people can do so with no costs, then suboptimal laws are no impediment to the economy functioning well; people just bargain around the suboptimal rules and the economy returns to the optimal. Thus, in a world where bargaining around rules faces no obstacles, distortions, or impediments, the law does not matter. The outcomes are identical regardless of the law.

This notion, that in a perfect world the law is irrelevant, is often called the Coase theorem. Coase did not express it as a theorem. Rather, Coase focused on transaction costs as an important impediment to people bargaining around the rules. If reaching an agreement to reallocate

© The Author(s) 2018
N. L. Georgakopoulos, *Illustrating Finance Policy with* Mathematica,
Quantitative Perspectives on Behavioral Economics and Finance,
https://doi.org/10.1007/978-3-319-95372-4_1

rights is costly, then individuals do not enter into the agreements that cancel any misallocations. Then, the legal rule needs to be correct. Thus, Coase justified some rules by building the much more important foundation that every rule needs to be justified. Each rule must help people overcome an impediment against allocating rights and obligations freely and, therefore, optimally.

A lot more analysis has dealt with this issue since Coase's article. Coase earned the 1991 Nobel Memorial Prize in Economics for his contribution and proceeded to lament the focus on his foundation—that the law is irrelevant—rather than that transaction costs justify laws, and to also lament the mathematization of economic analysis.[1] This chapter visits a couple of paradigmatic examples of Coasean irrelevance and its failures, which make the law relevant again. Readers who seek a fuller analysis may turn to an earlier book of mine.[2]

# 1    The Farmer–Rancher Example

The farmer–rancher example illustrates that the law is not relevant by considering a rancher who would save by having the cattle traverse a farm whereas the farmer would suffer from the cattle trampling the crops. The law at issue is whether the farmer is entitled to fences preventing the cattle from crossing compared to an "open range" rule prohibiting fences. If the farmer and the rancher can bargain without impediments, then the optimal outcome arises regardless of the rule.

The setting may be that the farm sits between a watering hole and the ranch. Driving the cattle around the farm to the watering hole costs an extra amount to the rancher, $-w$. Crossing the farm means that the rancher does not incur that cost but the farmer incurs the cost of trampling, $-t$. The optimal result is what would obtain if a single person

---

[1] *See, e.g.*, Brian Doherty, *RIP Ronald Coase, the Economist Who Hated Math*, REASON (Dec. 2013), available at http://reason.com/archives/2013/11/10/rip-ronald-coase-the-economist [perma. cc/93SG-7UGA].

[2] Nicholas L. Georgakopoulos, PRINCIPLES AND METHODS OF LAW AND ECONOMICS (2005) (devoting two chapters to Coasean irrelevance).

both farmed the farm and ranched at the ranch. This single owner would compare the cost of trampling to the cost of driving the cattle to the water the long way, and choose the least costly. Namely, the single owner would drive the cattle the long way if $-w > -t$ and would have the cattle cross and trample if $-t > -w$. For example, if the long drive costs $50 and the trample costs $100, the single owner would choose the long drive.

Suppose that the law imposes the wrong rule. The rule is that ranchers are entitled to an open range whereas the long drive costs less than trampling. Is the farmer condemned to suffer trampling? Coase argues that the farmer would bargain with the rancher and have the rancher take the long drive. If the farmer suffers $100 from trampling and the rancher suffers $50 from the long drive, the farmer is willing to pay any amount up to $100 to avoid trampling and the rancher is willing to accept any amount over $50 for taking the long drive. The two sides have room to reach an agreement for the rancher to not cross the farm.

Vice versa, suppose the rule is that the farmer has the right to fence the land but only suffers $25 from trampling while the long drive still costs $50 to the rancher. Will the rancher take the wasteful long drive? Again, Coase's analysis says no. The rancher is willing to pay up to $50 to avoid the long drive. The farmer should accept anything over $25 to suffer the trampling. The two sides have room to reach an agreement for the rancher to cross the farm, avoiding the long drive.

The point is that the parties circumvent the suboptimal allocation of the right by the law. The private initiative of the two sides leads them to bargain. The bargain leads to the optimal outcome, what a single owner of both the farm and the ranch would have done. The false law is irrelevant in the sense that the law has not changed the actual conduct.

## 2  The Polluter–Neighbor Example

The polluter–neighbor example takes us to a setting where the polluter sets up a factory that produces a harmful emission and a single neighbor suffers all the harm from that emission. Again, Coase argues that the sides will bargain depending on the cost of avoiding the emission and the harm to the neighbor. Whether the law requires the factory to have a filter that eliminates the emissions or allows the factory to pollute will not change conduct compared to what would occur with a single owner of both the factory and the neighboring property. Analytically, the setting is identical to the farmer–rancher example.

Suppose the filter costs $50 per month and the neighbor suffers $100 per month. A single owner of both would install the filter. Even if the law did not require the filter, the two sides would reach the same outcome. The neighbor would be willing to pay up to $100 to avoid the harm from the unfiltered emissions. The factory should be willing to install the filter if it were to receive anything over $50. The two sides have room for an agreement that would circumvent the suboptimal allocation of the right by the law.

Vice versa, if the neighbor's harm were only $25 per month, then a single owner of both would not install the filter. Even if the law did require the filter, the owner of the factory would offer up to $50 for the neighbor to suffer the harm without asking for a filter. The neighbor should accept offers above $25 to bear the damage. The two sides have room for an agreement that would circumvent the suboptimal allocation of the right by the law.

## 3    Impediments to Irrelevance: Transaction Costs and Others

The polluter–neighbor example also serves as an illustration of the importance of transaction costs. A transaction cost is a cost that impedes the reaching of an agreement (as opposed to the costs of fulfilling the agreement). Thus, the cost of traveling to the market, for example, is a transaction cost. The cost of providing a service to customers once at the market is not a transaction cost. Rather, one may call it a fulfillment cost, the cost of fulfilling the bargain to provide them with that service.

To see the importance of transaction costs, suppose that instead of a single neighbor, the polluter has 1000 neighbors who suffer from the pollution. Each suffers a thousandth of the harm but faces some cost in communicating with their neighbors. Suppose that one of the neighbors needs to mail the agreement to each one to sign at the cost of a postage stamp, rounded to $0.50. One thousand mailings produce a cost of $500. Using the numbers of the example, the agreement is not worth the mailing costs. That the mailing costs are greater than the gains of the agreement means that the parties will not enter in the agreement. Any false allocation of the right by the legal system will persist. Transaction costs can cause the failure of Coasean irrelevance.

Any impediment to reaching agreements can cause the failure of irrelevance. Transaction costs are not the only possible impediment to reaching an agreement.

The combination of the irrelevance of the law when agreements have no impediments and the existence of impediments is crucial for policy analysis. Once policy analysis accepts these premises, then the desirability of a certain outcome, such as environmental cleanliness, alone, is not grounds for regulation. Rather, for regulation to be justified, one must also identify an impediment to the private resolution of the matter or regulatory interest. In this way, the adoption of Coase's premise has made legal analysis much more rigorous and difficult. No longer is the desirability of an outcome a sufficient justification for law. For regulation to be justified, one must also show that the private initiative cannot reach that outcome despite its desirability. Sometimes this second step is trivially easy, as in the case of environmental regulation. Very often, however, establishing that an impediment to private solutions exists is the more difficult step of the analysis.

Returning to the impediments against private resolution of the allocation of rights, scholarship seems to have settled into accepting a few broad categories besides transaction costs: distribution of wealth, systematic errors, negotiation holdouts, and aversion to risk. These are not closed categories. Settings may exist with unique impediments and additional categories may emerge.

## 3.1    Distribution of Wealth

The most obvious impediment is distributional concerns. To whom a right is allocated determines who will receive compensation for parting with that right, in the cases when the right is not allocated optimally, and who will avoid expenditure, in the cases where the right is correctly allocated. Before celebrating the ease with which distribution may justify rules, however, one must be attentive to the analysis that rules that have distributional consequences distort incentives. The outcome for society is that redistributing through rules is usually inferior to redistributing through an optimal tax regime and letting the rules be optimal.[3]

## 3.2    Systematic Errors

One of the most popular sources of failures of Coasean irrelevance is systematic errors. In the economic literature, it appears under the rubric behavioral economics. The 2002 Nobel Memorial Prize in Economics as well as the 2017 prize decorated leaders in behavioral economics, Daniel Kahneman in 2002 and Richard Thaler in 2017. The relevance of systematic errors for failures of Coasean irrelevance is straightforward. If individuals tend to make certain mistakes, then their related decisions to bargain around rules will also tend to be erroneous. Therefore, those decisions may not promote social welfare. Intervention by the legal system may be, then, justified. The caveat in this category is that a strong reaction by the legal system, in the form of a mandatory rule, could end up hurting the very people who do not make the mistake. Professors Sunstein and Thaler have promoted, as a response, the principle of *libertarian paternalism*: rules designed to overcome the

---

[3]*See* Louis Kaplow & Steven Shavell, Fairness versus Welfare (2002); Nicholas L. Georgakopoulos, Principles and Methods of Law and Economics 84–89 (2005) (explaining the argument); Nicholas L. Georgakopoulos, *Exploring the Shavellian Boundary: Violations from Judgment-Proofing, Minority Rights, and Signaling*, 3 J. L. Econ. Policy 47 (2006) (expanding on the double distortion argument by identifying instances of rules that have both tax-reducing and redistributive appeal); Nicholas L. Georgakopoulos, *Solutions to the Intractability of Distributional Concerns*, 33 Rutgers L. Rev. 279 (2002).

biases that individuals tend to display but also designed so that individuals can elect not to be bound by these rules.[4] A new and little understood concern, about the discount rates for evaluating present costs against distant-in-time gains, also seems as a systematic error. Using falsely low discount rates will cause an expenditure that is not truly desirable because the future gain it produces, discounted correctly, does not justify the expense. A further caveat is that the evidence of systematic errors is not as widely accepted as one might like, with critics pointing out that the evidence stems from experimental settings that may not translate to real life.[5]

## 3.3    Negotiation Holdouts

A category of failures of Coasean irrelevance that has not been particularly fruitful for academic research is systematic negotiation imbalances or holdouts. Returning to the farmer–rancher example, suppose farmers systematically play hardball in the negotiation with the ranchers, meaning that they insist on a disproportionate share of the gains and allow some deals not to occur in order to achieve that.

For example, the right to fence is allocated to the farmer but crossing the farm is cheaper because it harms the farmer $25 whereas the long drive costs the rancher $50. An even division would divide the gain and have the rancher pay $37.50, the midpoint between $25 and $50. The rancher plays hardball and is only willing to pay $26, accepting the possibility that the farmer may reject the deal. This strategy might make sense for the rancher if the rancher repeats this game with many farmers and knows that only 5% of the deals are lost that way. Then, out of 100 deals the rancher will suffer cost of $50 on each of the failed deals (total cost of $250, i.e., 5 times $50) and only pay $26 on each of the

---

[4]*See generally*, RICHARD H. THALER & CASS R. SUNSTEIN, NUDGE: IMPROVING DECISIONS ABOUT HEALTH, WEALTH, AND HAPPINESS (2008).

[5]*See, e.g.*, Gregory Mitchell, *Why Law and Economics' Perfect Rationality Should Not Be Traded for Behavioral Law and Economics' Equal Incompetence*, 91 GEO. L.J. 67 (2002); Richard A. Posner, *Rational Choice, Behavioral Economics, and the Law*, 50 STAN. L. REV. 1551 (1998) (rebutting implications for law from the limitations of cognition).

95 other deals (total cost of $2470, i.e., 95 times $26) for a total cost of $2720. Consider the alternative of having all deals occur at the midpoint, $37.50. Having all the deals occur at $37.50 is worse, since then the total cost is $3750. Therefore, the holdout strategy makes sense for the rancher. However, for the entire economy, this is undesirable. Total productivity suffers the loss of the missed deals.

The caveat with negotiation holdouts as a source of failures of Coasean irrelevance is the limited understanding of their circumstances. Because they are private bargains, no data on their structure exists.

## 3.4    Aversion to Risk

A very different source of failure of Coasean irrelevance is risk aversion, because it is a common trait, it is ubiquitous. As we will see in the chapter on aversion to risk (Chapter 10), when an individual faces an uncertain outcome, the usual risk-averse individual would accept a smaller certain outcome (the certainty-equivalent of the risk). Applied to Coasean bargains, this means that if one course of action entails a risk, then individuals will not assess it neutrally but will see it through the prism of their aversion to risk. Individuals will discount the uncertain outcome. Then, if the superior outcome involves a risk, individuals will undervalue that and some bargains may not be struck or suboptimal bargains will be struck. The entire economy, society as a whole, is much closer to being neutral toward risk, because it is diversified. The entire economy contains the many individuals who will tend to experience the probabilistic outcome in the proportions that correspond to the probability of the outcomes.

For example, the rancher's cost of the long drive is $50 but only on average. The probabilistic outcome is that half of the time the cost of the drive will only be $5 because the cattle will be well behaved; the other half of the time, however, the cattle will be ornery, requiring much extra effort, making the drive cost $95. The rancher may be sufficiently averse toward risk to consider the $5-or-$95 risk as being equivalent to a loss of $80.[6] Then the rancher will bargain to cross the

---

[6]In the chapter on aversion to risk, Exercise 5 in Chapter 10 asks you to calculate these ranchers' coefficient of risk aversion.

farm even if the cost of crossing is, say, $75. Thus, the costlier action occurs, the cost of $75 from the trampling of the farm instead of the cost of $50 on average from the long drive. A society composed of many ranchers and farmers would rather have them all take the long drive and half regret it rather than all reach the agreement with the farmers to cross the farm.

The caveat to embracing risk aversion as a reason for the failure of Coasean irrelevance is that private initiative can perceive the problems posed by risk and can design effective ways to overcome risk aversion. Examples are the development of insurance, and of cooperatives and other mechanisms of making the danger common to a pool (such as partnerships and corporations or any business association). A regulatory resolution of a problem may displace a more cost-effective solution from private initiative.

## 4    Conclusion

The concept that the bargains that individuals strike are not just good for the parties but tend to also advance aggregate economic welfare is one of the key contributions of economic analysis. Its extension to law with Coasean irrelevance reduces the tendency to see law with a paternalistic attitude but makes policy analysis much more difficult. For policy analysis to proceed after Coase, one who proposes a legal intervention must also demonstrate that private bargaining cannot reach the optimal outcome. Several sources of failure exist, restoring the desirability of law as a mechanism that will bring optimal conduct to society.

## 5    Exercise

1. Consider the election of representatives to a legislative body as a means to overcome the costs for a large number of members of society organizing and becoming represented to reach a Coasean bargain opposite a unified entity. Using the polluter–neighbor example,

electing representatives can be seen as a mechanism for the many neighbors to organize so that their representatives can bargain with the polluting factory. How would such a vision of democratic institutions inform their design? Consider eminent domain law (the power of the government to expropriate property), centralization opposite decentralization or local authority (and all aspects of federalism), as well as any other features of democratic regimes you wish.

# 2

# Introduction to Mathematica:
# Hello World in Text and Graphics

Instructional manuals for programming languages usually start with the project of writing the code to display "Hello World" on the screen. This chapter introduces the basics of Mathematica and displays "Hello World" on the screen, which is trivial, but also in a graph.

## 1    Kernel, Notebook, and Cells

Mathematica consists of an invisible program called the kernel that does the computations and a visible program called the front end that reads the user's instructions and displays the output that the kernel produces. Thus, to send commands to Mathematica, we the users type them in a document created using the front end.

The documents that users create with the front end are notebooks. Thus, the new blank document that you see when you open Mathematica is a notebook.

When you start typing in the notebook, a thin blue square bracket appears on the right margin. If you hit enter and continue typing, the

© The Author(s) 2018
N. L. Georgakopoulos, *Illustrating Finance Policy with* Mathematica,
Quantitative Perspectives on Behavioral Economics and Finance,
https://doi.org/10.1007/978-3-319-95372-4_2

bracket will expand to both lines. That bracket identifies a cell of the notebook. Mathematica will ask the kernel to evaluate the cell when you hit Shift-Enter.

Cells come in several varieties. The cell you just created is an input cell. Mathematica will take it as commands for (i.e., as input for) its computing algorithms in the kernel. The cell(s) that Mathematica creates in response are output cells. Those are the two main types of cells. If you do nothing but start typing, Mathematica puts what you type into an input cell. If you start typing in an output cell, Mathematica will convert it into an input cell to include your typing.

## 2    More About Cells

Three other cell types might prove useful. If before you start typing you hit Alt-1, then Mathematica converts the cell that is about to be created into a title cell. Hitting Alt-4 creates a section cell, and Alt-7 creates a text cell. The intervening values (Alt-2, Alt-3, etc.) are subtitles and subsections. The reason you might want to use sections is that they are collapsible. I often create a single notebook for a large project. After several weeks, the notebook has a title, an initialization section, and several sections with the various analyses. If I remember to double-click the bracket on the right of each section before I save, then when I open the notebook, all the sections are collapsed and I can expand and work on only the one I want, without trying to find my place in a sea of code and computations.

The reason that you likely want to use text cells is to prevent Mathematica from evaluating them. If Mathematica evaluates plain text, it will likely produce an error or jumble the words in its reaction (i.e., in the output cell that Mathematica will create when trying to evaluate the input). That occurs because Mathematica will treat each word in an input cell as a command or a variable, and it will follow its rules about formatting mathematical output in deciding how to show it. Usually, this means putting the words (which Mathematica thinks are variables) in alphabetical order.

# 3 Simple Commands on the Screen

Having understood the basics, let us send Mathematica some commands. Open a new document. Type $2 + 2$, hit enter, and type Hello World without following it with a period or anything else. Then hit Shift-Enter to send those two commands to Mathematica. Your screen should look like Fig. 1.

In reaction to our command, Mathematica did a lot, so we will have to take it a step at a time. First, Mathematica took our commands and created two output cells. The first one holds the number 4 as the result of the arithmetic operation 2 plus 2. Mathematica created a second output cell saying "Hello World" in response to our second command.

Beyond these relative straightforward reactions, notice that Mathematica displays our Hello World input in blue instead of black. This shows that Mathematica did not understand what Hello and what World stand for. If we intended them as commands, Mathematica does not recognize them. If we intended them as variables, again, Mathematica does not recognize them because we have not defined them. Mathematica does not treat them as a string of text because they are not inside quotation marks.

**Fig. 1** Our first commands

At the right end of the window, immediately next to the scroll bar, Mathematica displays some narrow blue vertical brackets in two columns. The three brackets on the left identify the cells: our two-line input cell, and the two single-line output cells. The rightmost single gle bracket corresponds to the grouping of the three cells into a single gle group of matched input and output. If we double-click there, Mathematica will collapse the grouped cells and only show the topmost cell. We can select one or more of those brackets with the mouse to copy, cut, or evaluate an input cell again.

Also, Mathematica puts "In[1]:=", "Out[1]=", and "Out[2]=" in front of the three cells. Those correspond to Mathematica's numbering of its inputs and outputs. We can use this numbering to bring them back later. In the meanwhile, Mathematica remembers every input and output of each session. If we ask Mathematica to execute the same input cell again, Mathematica will treat that as a new input, since we may have changed the cell in the meanwhile. If we save this document, exit Mathematica (or quit the kernel through the command bar under Evaluate/Quit Kernel), and reopen the same document, then Mathematica will begin a new session with new numbering. But if we have two documents open and evaluate cells in both, back and forth, all the inputs of the two documents will form a single sequence of inputs to Mathematica's kernel, and the outputs will also form a single sequence. Thus, Mathematica documents are unlike spreadsheets. If we have two open notebooks and define a variable in one, then use the same variable in the other, Mathematica will use the variable's prior definition. While that sounds counterintuitive and undesirable, it has benefits too. If you are making a presentation or otherwise showing something, you do not need to clutter the demonstration with definitions. You can execute all the cells in a different notebook for Mathematica to load all the definitions. Then you can turn to the cleaner demonstration notebook and bypass the clutter.

Immediately below all of the above, Mathematica may offer a gray line with buttons holding Mathematica's guesses of what we might want to do next, such as produce a 3D plot or a derivative of Hello. This book will never use this feature, and you can avoid the clutter by closing it from the small circle with an $x$ at its right end. A little gray circle with an arrow will appear there and lets you bring this bar of suggestions back.

The title line of the window has also changed from "Untitled-1" to "Untitled-1\*" to show that we have not saved this notebook in its current state. The title line continues to read "-Wolfram Mathematica" followed by the version you have. I have hidden the version here, since the front end seems to have reached a steady state without big changes from version to version. Older versions of Mathematica keep the menu bar separate from the notebook, so that the menu bar is at the top of the screen, akin to a menu bar on an Apple Macintosh computer.

## 4     Lists and Their Basics

That we made Mathematica display "Hello World" in this fashion, as in Fig. 1, is unremarkable and wrong, because Mathematica displays it as undefined variables. We will use "Hello World" to introduce a fundamental concept that Mathematica uses in its operation. Mathematica relies on lists.

Lists in Mathematica go inside curly brackets, i.e., { and }. Commas separate the elements of lists. (A space makes Mathematica try to multiply what it finds on either side of it, following scientific notation. However, when nothing is on one side of a space, then Mathematica will ignore the space, as in the list below.) Thus, we make a list with the letters of "Hello World" by creating a list of each letter inside quotes. My screen now looks like Fig. 2. I create a variable named helloworld. I assign to it a value using the equal sign. The value I assign to the variable is a list. The list is inside curly brackets. The content of the list is a sequence of letters inside quotation marks separated by commas.

```
In[3]:= helloworld = {"H", "e", "l", "l", "o", " ",
        "W", "o", "r", "l", "d"}

Out[3]= {H, e, l, l, o,  , W, o, r, l, d}
```

**Fig. 2**  Creating a list with the letters of Hello World

The letters make up "Hello World" including the space but no punctuation marks. When I send this command to Mathematica, Mathematica responds by displaying the list. I could have placed a semicolon at the end of the line to suppress the output.

By convention, variables are in lowercase. Mathematica is sensitive to capitalization. The internal commands of Mathematica use initial capitals. Therefore, we will avoid using initial capitals when naming. Accordingly, we named our list helloworld rather than Helloworld.

Lists are powerful. Mathematica can treat lists as vectors or matrices and apply matrix algebra to them, but this book will have no matrix algebra. As this book focuses on graphics, we will mostly use lists to store data and retrieve the data for graphing.

We can take individual specified members of a list using the **Part[ ]** command or placing the number of the member in double square brackets after the name of the list. **Part[helloworld, 2]** and **helloworld[[2]]** are equivalent and their output is **e**, the second member of the list.

Mathematica sends lists to commands in four important ways, three built-in methods and one customizable method. The built-in methods are (1) directly as lists, (2) by using **Apply[ ]** or **@@**, and (3) by using **Map[ ]** or **/@**. The customizable way is to design pure functions by using **Function[{symbols}, commands]** or **commands[#] &**. Figure 3 illustrates this.

The straightforward way of applying a command to a list is to pass the list to the command by putting it in square brackets following the command or connecting them with **@**. Mathematica also lets us place the command after its argument if they connect with a double slash, **//**. As a result, the following three are equivalent: **Print[h]**, **Print@h**, **h//Print**. In the context of our example, if we pass the helloworld list to the **Print[ ]** command, Mathematica will print the list, including the curly brackets and commas.

The second method is to use the **Apply[ ]** command (or **@@**). Then Mathematica effectively removes the curly brackets and replaces them with the command. In the context of our example, if we use **Apply[ ]** to pass the list to the **Print[ ]** command, we get the printed text Hello World. Granted, this is not economical. It would

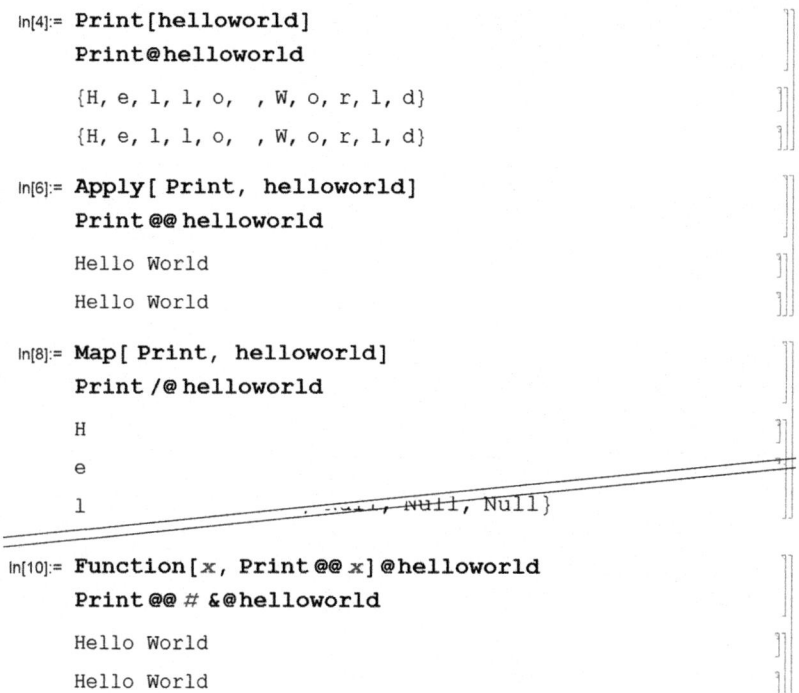

```
In[4]:=  Print[helloworld]
         Print@helloworld

         {H, e, l, l, o,  , W, o, r, l, d}
         {H, e, l, l, o,  , W, o, r, l, d}

In[6]:=  Apply[ Print, helloworld]
         Print @@ helloworld

         Hello World
         Hello World

In[8]:=  Map[ Print, helloworld]
         Print /@ helloworld

         H
         e
         l              ..., Null, Null}

In[10]:= Function[x, Print @@ x] @helloworld
         Print @@ # &@helloworld

         Hello World
         Hello World
```

**Fig. 3**  Four ways of passing a list to a command

have been much simpler to send a single string with all the letters to the **Print[ ]** command.

The third approach is to "map" the command over the list by using **Map[ ]** or **/@**. Then the command applies separately to each individual member of the list. When we map **Print[ ]** over our list, Mathematica produces an output cell printing each letter (and, at the end, a new list containing the null result of each application of the command, a list of **Null**s).

The fourth approach is to create what Mathematica calls a pure function. Pure functions, despite being defined by the user, are not the kind of user-defined functions that offer the main power of Mathematica programming, which we will encounter often, and which are introduced at p. 29. Pure functions are very brief definitions of functions that have a slot or more for variables following the pure function.

The slot is represented by a hash mark, **#**, or a hash mark followed by a number if the pure function uses several variables. The abbreviated definition of a pure function ends with the ampersand, **&**. By contrast, the main user-defined functions can be very long, and their definition comes well before the variables that those functions will use.

# 5    *Table[ ]*

A simple way to deal with lists for graphing is to use the **Table[ ]** command. The **Table[ ]** command is a looping programming command; its output is a list. The syntax of the **Table[ ]** command is **Table**[command, {iterator, *minimum*, maximum, *step*}]. An example is **Table[i, {i, 3, 15, 6}]**, which will create the list {3, 9, 15}. In the example the iterator was *i*, with minimum 3, maximum 15, and step 6. The command was to output the value of the iterator *i*. The italicized components are optional. If either is missing, then Mathematica assumes it takes the value one. The **Table[ ]** command can take a list of commands, and several **Table[ ]** commands can be nested.

As an example, let us use **Table[ ]** to place the letters of the list helloworld on a sine curve in a two-dimensional graph. To show a two-dimensional graph, we send Mathematica the command **Graphics[ ]** or **Show@Graphics[ ]**. Inside it we place a list of graphics commands that Mathematica calls graphics primitives, meaning basic graphing commands. The graphics primitives may be shapes, lines, colors, etc. The graphics primitive we will use is **Text[ ]** and its syntax is **Text**[textstring, {x-coordinate, y-coordinate}]. We will apply the **Text[ ]** command to each letter, we will increase the horizontal x-coordinate gradually, and we will let the vertical y-coordinate take values from the sine function, **Sin[ ]**. Figure 4 is a screen capture of the result, omitting a few lines of intermediate output.

Again, a lot happened that needs explanation. First, we defined a new variable, **st1**, to hold the result of the **Table[ ]**. The command itself takes four lines. The first line merely begins the definition of the new variable as equal to the output of **Table[ ]**. The iterator is at the fourth line. The name of the iterator is i, the minimum is defined to be one (unnecessarily so, since it would be one if omitted). The maximum

```
st1 = Table[

  Text[helloworld[[i]] ,

    {i, Sin[ ——————————— ]}}]
           Length[helloworld]
            i 2 π

  , {i, 1, Length[helloworld]}]
```

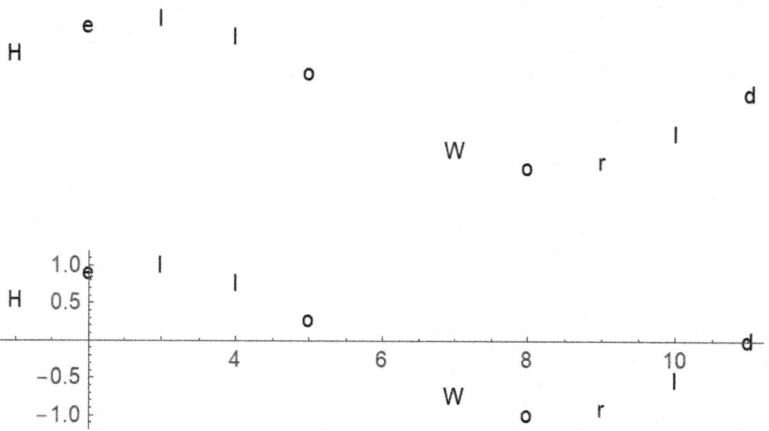

```
Graphics@st1
Graphics[st1, Axes → True]
```

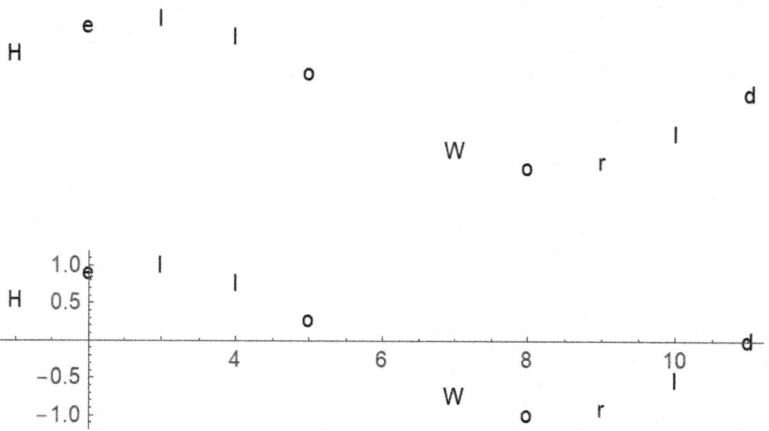

**Fig. 4**  Using Table[ ] to place letters in a graph

is the length of the list helloworld. The omission of a step makes the step one. The only command inside the **Table[ ]** is **Text[ ]**, which occupies lines two and three. Line two starts by stating the **Text[ ]** command and having it apply to the *i*th element of the list helloworld. Instead of **Part**[helloworld,  i], the code uses its double-square-bracket abbreviation: **helloworld[[i]]**. Line three establishes the coordinates. The *x*-coordinate is i. The *y*-coordinate is (i2$\pi$)/(length of helloworld), explained further below. The way that $\pi$ gets represented could have been **Pi**. By using escape-p-escape, the Greek letter gets placed in the notebook and Mathematica understands it as the pi constant. The way to create a fraction is Ctrl-/ or to use one of the palettes that Mathematica offers that have buttons for inserting various mathematical symbols.

The result of this **Table[ ]** is a list, a sequence of **Text[ ]** commands containing each letter in the list and its coordinates. To shorten the picture, the figure omits the middle of the output. Then **Graphics@st1** displays the graph. The same graph but with axes comes from **Graphics[st1, Axes->True]**. In the language of Mathematica programming, the components of the **Graphics[ ]** command that come after the list of graphics primitives are options. Options change the defaults that Mathematica uses in various commands. Here we changed the option about displaying axes from its default of not showing the axes to showing them.

A further explanation is necessary for the construction of the *y*-coordinate, namely Sin[($i2\pi$)/(length of helloworld)]. The sine function has the form of a wave. The function starts at zero, crests at the value of one when the input is $\pi/2$, passes by zero at $\pi$, troughs at $-1$ when the input is $3\pi/2$, and returns to zero at $2\pi$. In other words, Sin[0] is zero, Sin[$\pi/2$] is one, Sin[$\pi$] is zero, Sin[$3\pi/2$] is minus one, and Sin[$2\pi$] is zero. The letters of helloworld follow this wave. Several approaches could produce the same outcome. One might choose to let the iterator take values in that range and design the selection of the letter from the list **helloworld** through some transformation that would produce integer numbers. The method chosen here is to let the iterator run the length of the list but scale it for the purpose of the sine function. The scaling is such that at its maximum, the length of the list, the resulting value is $2\pi$. Therefore the scaling factor becomes $2\pi$/(length of the list). When this is multiplied by the maximum value that the iterator *i* takes, the result is $2\pi$. The smaller values of the iterator gradually approach that value. The result is that the letters of the list take their places along one complete period of the sine curve.

Most of the above discussion assumed a simple usage of the various commands. Most of the commands that this chapter discussed have more components, a more complex syntax. The help files of Mathematica reveal this complexity.

# 6  Conclusion

In this chapter, you learned the basics of Mathematica. Mathematica documents are notebooks and contain cells of various types. You send commands to Mathematica by writing input cells and hitting Shift-Enter. Notebooks are not independent: a single Mathematica session may use several notebooks. Mathematica's built-in commands use initial capitals (and we, to avoid confusion, use lowercase for naming what we define). Mathematica uses lists and applies commands to lists in several ways. A basic looping command is **Table[ ]**. We saw the use of **Table[ ]** to create and display some graphics primitives in a two-dimensional graph using **Graphics[ ]**. While the built-in plotting commands of Mathematica are powerful, building graphs from graphics primitives allows much greater customization.

# 7  Exercises

1. Use the built-in help of Mathematica to determine additional potential components of the **Text[ ]** command. What are they?
2. The discussion of ways to pass a list to a function omits the temporary assignment of a value to a symbol using slash-dot ("/."). Create an example using this way to pass a value to a function.
3. Observe that in the sine graph of hello world, the $y$-coordinate of the first letter, H, is not zero. Instead, the H floats above the horizontal axis. Why is that? How can you change the **Table[ ]** command that produces the list **st1** so that the letters still follow the sine curve but start and end at a $y$-coordinate of zero?
4. Consider    the    command    **Point@Table[{j, Cos[j]}, {j,0,10,.2}]**. What does it do? Can it immediately follow

**Graphics@**? Does mapping **Point[ ]** on the same **Table[ ]** command work? How would you write a pure function to get individual pairs of coordinates from the table inside each **Point[ ]** command so that graphing one element of the resulting list would graph one point?

5. Adapt the table of the exercise about placing hello world on a sine wave to three dimensions, so that using **Graphics3D** on it would display a helix in three dimensions.

# 3

# The Mathematical Frontier: Trigonometry, Derivatives, Optima, Differential Equations

Graphical applications bring up trigonometrical functions, such as sine and cosine, frustratingly often. This chapter uses Mathematica's graphics to review the basics of trigonometry. Being in a review mode, the chapter continues to review derivatives. We close with our first legal application, differential equations applied to legal change.

## 1    Trigonometry Illustrated

To illustrate trigonometry, begin by creating a circle with radius one and with its center at the origin of the coordinate system, i.e., at the point (0, 0). Use Mathematica's `Circle[]` command. Its syntax is `Circle[{`x-coord, y-coord`}, `*radius*`]` but radius is set at one if omitted. `Graphics[Circle[{0,0}], Axes->True]` produces the desirable outcome.

© The Author(s) 2018
N. L. Georgakopoulos, *Illustrating Finance Policy with* Mathematica,
Quantitative Perspectives on Behavioral Economics and Finance,
https://doi.org/10.1007/978-3-319-95372-4_3

## 1.1    An Angle, Its Sine, and Cosine

Form an angle using the horizontal axis, the x-axis, and a line that intersects the axis at the origin. Measure the angle counterclockwise from the positive values of the x-axis. The sine of this angle is the height, or *y*-coordinate, of the point where the line that intersects the axis crosses the circle. The cosine of the same angle is the *x*-coordinate of the same point. Thus, we can place this point by using the **Point[ ]** command. **Point[{Cos**[angle]**, Sin**[angle]**}]** would produce that result. To emphasize the point, increase its size by preceding it with a **PointSize[ ]** command. Form the angle by drawing the line that intersects the x-axis. Figure 1 shows my resulting screen.

Several notables: (1) All the graphics primitives are in a list, inside curly brackets. (2) Rather than beginning directly with the **Graphics**[ ] command, the corresponding line first assigns the graphic to a variable, **uc1**. (3) Mathematica measures angles using radians, which use $\pi$, so that a 90-degree angle is $\pi/2$, a 180-degree angle is $\pi$, and an angle that coincides with the circle is $2\pi$. The constant **Degree** converts angles from degrees into radians. Nevertheless, despite that this makes using angles expressed in degrees easy, using $\pi$ produces consistency, because some output will be expressed in terms of $\pi$. Getting used to the idea that a circle is $2\pi$ would be great. (4) The code defines the angle in a single point, before starting the code for the graphic, by placing its value in the variable **rads**. This way all functions inside the graphic can refer back to that variable. This is a very convenient programming habit. Any value that appears several times deserves a separate definition. Then, a single edit will change the variable in every place it arises. In this example we are building, the separate definition of the angle will let us eventually animate the graphic. (5) The arc near the origin that defines the angle uses a more complicated syntax of **Circle[ ]**. This syntax includes a third element after the circle's center and radius, the beginning and end of the arc. The syntax is **Circle[{**x-coordinate of center**,** y-coordinate of center**}**, radius**,** {angle of beginning of arc**,** angle of end of arc**}]** and,

```
rads = .3;
uc1 = Show @ Graphics [{
    Circle [{0, 0}],  (*unit circle*)
    Circle [{0, 0}, .15, {0, rads π}],  (*angle*)
    PointSize [.02],
    Point [{Cos [rads π], Sin [rads π]}],  (*point on unit circle*)
    Line [{{0, 0}, {Cos [rads π], Sin [rads π]}}] (*line to circle*)
    }, Axes → True]
```

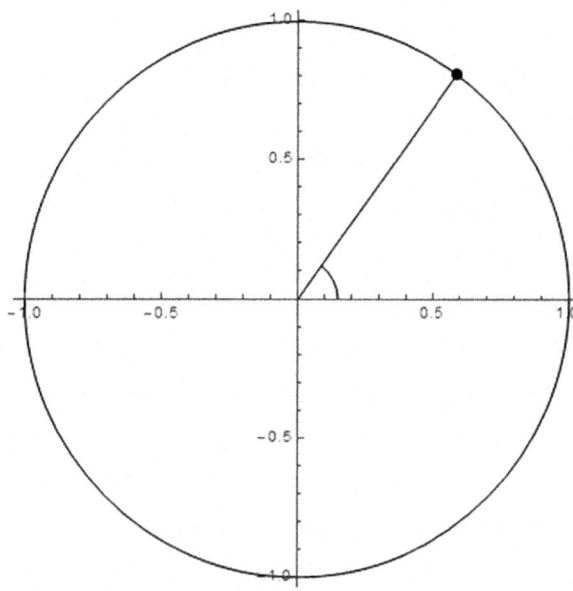

**Fig. 1**  An angle on the unit circle

here, it takes the form **Circle[{0,0},.15,{0, rads π}]**. (6) Text inside parentheses with stars, i.e., **(\*** such as this text **\*)**, are ignored by Mathematica. These are remarks. Mathematica grays this text out. The code here and throughout this book will have voluminous remarks explaining how it works.

Figure 1 is still quite crude. Improve it by placing dashing trace lines from the point to each axis, by including the tangent and cotangent, and by labeling the trigonometric functions.

Since we already know the coordinates of the point on the circle, the location of the ends of the trace lines on the axes is trivial. Place the commands forming the lines at the end of the list of graphics primitives. Precede them with the **Dashing[ ]** command to make them dashing.

## 1.2    Tangent and Cotangent, *Solve[ ]*

The trigonometric function tangent gives the height (or *y*-coordinate) where the line forming the angle reaches a vertical line that is tangent to the unit circle at the point (1, 0). The corresponding command in Mathematica is **Tan[ ]**. Place a vertical line with an x-coordinate of one. Use the **Tan[ ]** function to place a point on that line.

Changing the angle (**rads**) will change the location of that point. The problem is that the point can rise above or drop below the ends of the tangent line. To avoid displaying the point away from the tangent line, use an **If[ ]** statement to only draw the point if it would fall on the tangent line. The code below will have the tangent line extend from *y*-coordinates of −1.5 to 1.5. To find when **Tan[ ]** would fall outside that range, use the **Solve[ ]** function to find the value for which **Tan[x π]** is equal to 1.5.

The syntax of the **Solve[ ]** command is **Solve[**left-hand side of equation==right-hand side of equation**,** variable for which to solve**]**. Accordingly, the needed command is **Solve[Tan[x π]==1.5, x]**. Mathematica produces an error because this equation has infinite solutions and Mathematica gives only the first. By understanding the periodic nature of the tangent function, that single solution can produce all the boundaries necessary for the **If[ ]** construction. Note that Mathematica uses the double equal sign for testing equivalence. The single equal sign assigns a value to a variable even inside a command like **Solve[ ]**.

The syntax of **If[ ]** is **If[**condition**,** commands to execute if the condition is true**,** *commands to execute if the condition is false***]**. The point will be on the line (in other words, **Tan[rads π]** will have a value between −1.5 and 1.5) if **rads**, first, is smaller than .312 (ignore negative values for this angle for now). The logic of trigonometry is that the point will also be on the line when the angle approaches the full circle and becomes greater than $2\pi$ − .312. The point will also be on the line when the angle is on the opposite side of the circle, between $\pi$ − .312 and $\pi$ + .312, but ignore this for now.

Thus, we need the **If [ ]** command to produce one set of statements if the angle is smaller than .312 or larger than $2\pi$ − .312. When the angle is outside those boundaries, i.e., the above are true, then the point on the tangent line is visible. Also, in those cases we should extend the line (the radius) that forms the angle to reach the tangent line. In the opposite case, when the angle is greater than .312 and smaller than $2\pi$ − .312, then we need only draw the line that defines the angle as a radius of the unit circle. Later, we will still draw the point on the tangent for values that take the angle to the opposite side of the circle.

The double condition means the **If [ ]** statement must either use the **Or [ ]** (abbreviated | |) command with the former way of stating the boundaries or the **And [ ]** (abbreviated **&&**) command with the latter statement of the boundaries. The code uses **&&**. The code below produces Fig. 2.

```
rads = .3;
uc1 = Show@Graphics [{
 Circle [{0, 0}], (*unit circle*)
 Circle [{0, 0}, .15, {0, rads π}], (*angle*)
 PointSize [.02],
 Point [{Cos [rads π], Sin [rads π]}], (*point on unit circle*)
 Point [{Cos [rads π], 0}], (*point on x-axis*)
 Point [{0, Sin [rads π]}], (*point on y-axis*)
 Line [{{1, -1.5}, {1, 1.5}}], (*draw line tangent to unit cir-
 cle at (1, 0)*)
  If [rads > .312 && rads < 2 - .312, (*range of angle not point-
  ing to the tangent line*)
  Line [{{0, 0}, {Cos [rads π], Sin [rads π]}}] (*only  graph  line  to
  unit circle, not to tang't*)
  (*place for if [] statement drawing point when angle on opposite
  side*)
  , (*else*)
```

```
{Line[{{0, 0}, {1,
  Tan[rads n]}}], (*graph line to tangent at one*)
Point[{1, Tan[rads n]}]}(*point on line tangent at one*)
], (*endif*)
Dashing[.02],
Line[{{Cos[rads n], 0}, {Cos[rads n], Sin[rads n]}}], (*draw dash-
ing sine trace*)
Line[{{0, Sin[rads n]}, {Cos[rads n],
  Sin[rads n]}}] (*draw cosine trace*)
}, Axes -> True]
```

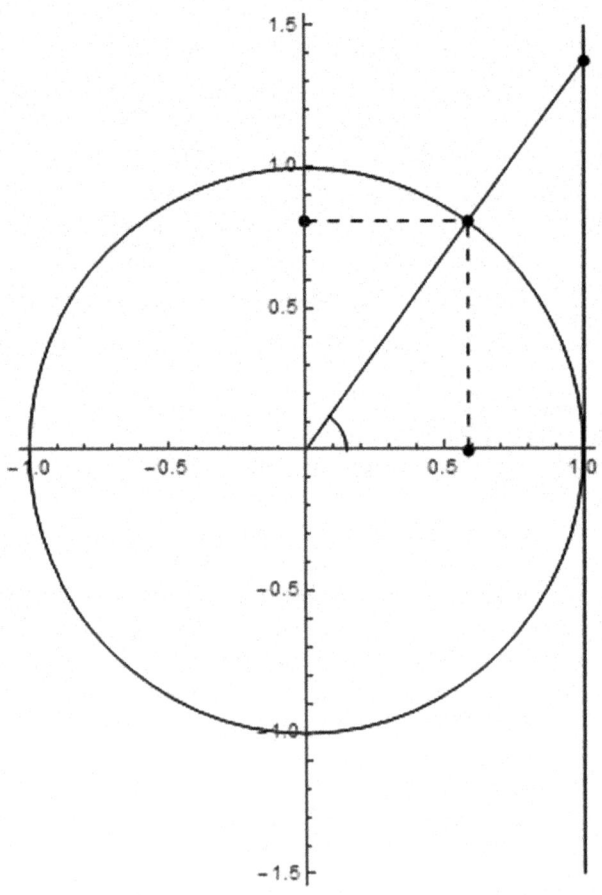

**Fig. 2** Adding dashing traces and the tangent

The cotangent is the *x*-coordinate of the point where the line that defines the angle crosses a horizontal line tangent at the unit circle at the point (0, 1). Set its boundaries at −1.2 and 1.2 and place the line. Solve the corresponding equation to find when the point is visible and formulate the corresponding **If [ ]** command.

## 1.3 Labels, Creating Functions

The **Text [ ]** function, already used in the first project of displaying "Hello World" on a sine wave above at p. 18, could be used with no change to place labels. However, that would confine us to the use of the default fonts and sizes. Mathematica offers customization for text through the **Style [ ]** command. Wrapping the text string inside a **Style [ ]** command allows control of the font and size of the text.

The control that **Style [ ]** offers comes at the cost of enormous verbosity. Many text labels will appear in graphs and having to use the nested **Text [Style [ ] ]** syntax will make the code very cumbersome.

Mathematica lets us circumvent this problem by defining a function. Notice that this is a long user-defined function rather than the short pure function that we saw in the basic ways of passing variables to a command above at p. 13.

Function definitions in Mathematica use several conventions. (1) Instead of the equal sign, function definitions precede it by a colon, **:=**. Mathematica literature calls this *delayed assignment*. By using **:=**, Mathematica does not evaluate the right-hand side of this assignment. The evaluation will come when the code invokes the left-hand side, likely also passing some specific values for the evaluation of the right-hand side.

(2) The function being defined can pass values to variables in its body (the right-hand side) by having them appear followed by underlines in the left-hand side. For example, we can define a function by **txt [a_ , b_ ] :=** followed by commands that use the variables **a** and **b**. When we later type **txt [5, 6]**, those specific values will be passed to the commands in the right-hand side of the definition of the function, and the result of the application of those commands will appear. In the case of

the **txt [ ]** function being designed here, the first variable would take the form of a text string. The second variable would take the form of a list with the coordinates for the text.

(3) However, if we define such a function with two arguments and we later invoke it with a different number of arguments, Mathematica will consider such an invocation as referring to an undefined function. This is especially problematic in the example of **Text [ ]**. Although **Text [ ]** is often invoked with only two arguments, the text string and its location on the graph, **Text [ ]** may also receive more arguments: A third argument determines text alignment, left, center, or right. A fourth argument determines the direction of the text. Therefore, a function that will pass arguments to a **Text [ ]** command needs to contend with the problem of sometimes receiving only two arguments from the code but sometimes receiving three or four arguments. The inelegant solution is to define the function several times, with a different number of arguments each time. Mathematica allows a more elegant solution: the use of optional arguments with default values for the case that the later code does not give corresponding arguments. Function definitions allow optional arguments by following the corresponding variable with colon and the default value. Thus, if we define the function **txt [a_ , b_ , c_ :**defaults**, d_ :**defaults**] :=***commands*, the later code can invoke **txt [ ]** with between two and four arguments, with the missing arguments taking the specified default values.

Perusing the Mathematica help files reveals that the default value for the third argument of the **Text [ ]** function is {0, 0}, which centers the text horizontally and vertically at the specified coordinates. The default value for the fourth argument is {1, 0} which produces horizontal text. The **Text [ ]** function interprets the argument about the direction of the text as a list with coordinates of the target direction. For example, {1, 1} would produce text sloping diagonally up and {0, −1} would produce vertical text reading downwards. The default values that Mathematica sets are the appropriate ones for the way that the custom function should behave but our code needs to restate them. Therefore, the left-hand side of the function definition should be **txt [a_ , b_ , c_ :{0,0}, d_ :{1,0}]**.

An appropriate right-hand side of the function defining how text should appear in graphics is `Text[Style[a, FontFamily->"Times", FontSize-> ftsz], b, c, d]`. The variable `ftsz` is one that could be defined universally for all the graphics in a notebook. Granted, the left-hand side of the definition could include it, letting subsequent code change it.

The variable with the size of the font and the definition of the function for displaying text in graphics can go in an initialization cell at the beginning of the notebook. The cell would have two lines:

```
ftsz=12;(*font size*)
txt[a_, b_, c_:{0, 0}, d_:{1, 0}]:= Text[ Style[a, FontFamily->"Times", FontSize -> ftsz], b, c, d]
```

Having defined the default size of the font and chosen a font for the text, subsequent code can use the `txt[ ]` function without having to repeat all its detail.

The simplest form of the `txt[ ]` function appears when the text is horizontal. The code to display "Cosine" at the center of the trace from the point on the circle to the y-axis (which is offset by **tol** above the axis) is

```
txt["Cosine", {Cos[rads π ]/2, Sin[rads π ] + tol}]
```

The $x$-coordinate of the text is the center of the trace line, half the value of $x$-coordinate of the point on the circle. The $y$-coordinate of the text is the $y$-coordinate of the point plus the variable where the text offset value is stored, **tol**. The code must define the variable **tol** earlier. Notice that all coordinates continue to depend on the angle, **rads**.

When the text must appear vertically, the fourth parameter of the `txt[ ]` command must be specified as {0, −1}. The code and the resulting graphic are as follows (Fig. 3).

```
rads = .3;
tol = .05;(*text offset from line below text*)
ucl = Show@Graphics[{
 Circle[{0,0}],(*unit circle*)
 Circle[{0,0}, .15, {0, rads π}],(*angle*)
```

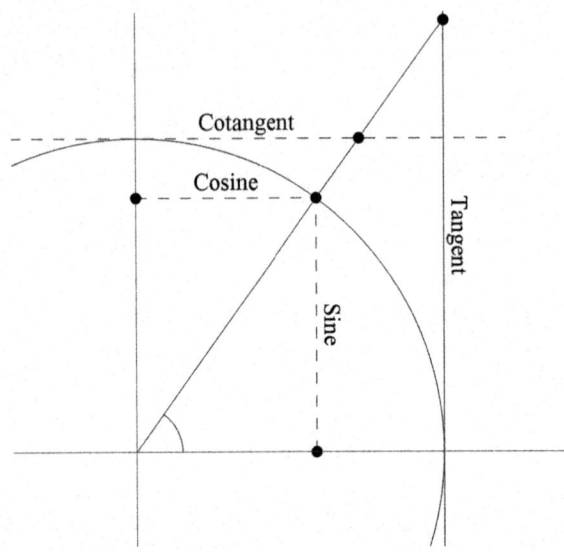

**Fig. 3**  Illustrating four trigonometric functions

```
PointSize[.02],
Point[{Cos[rads π], Sin[rads π]}], (*point on unit circle*)
Point[{Cos[rads π], 0}], (*point on x-axis*)
Point[{0, Sin[rads π]}], (*point on y-axis*)
If[rads > .312 && rads < 2 - .312, (*range for point on tangent
not visible*)
  Line[{{0, 0}, {Cos[rads π], Sin[rads π]}}] (*only graph line to
  unit circle*)
  , (*else*)
  {Line[{{0, 0}, {1, Tan[rads  π]}}], (*graph line to  tangent
  at one*)
  Point[{1, Tan[rads π]}], (*point on line tangent at one*)
  txt["Tangent", {1 + tol, Tan[rads π]/2}, {0,
  0}, {0, -1}]} (*and label*)
  ], (*end if*)
If[rads > .222 && rads < 1 - .222, (*boundaries for visible cotan-
gent point*)
  {Point[{Cot[rads π], 1}], (*draw point on cotangent at (0,1)*)
  txt["Cotangent", {Cot[rads π]/2, 1 + tol}]}
  ], (*end if*)
Line[{{1, -1.5}, {1,
  1.5}}], (*draw line tangent to unit circle at (1, 0)*)
```

```
Dashing[.02],
Line[{{-1.2,1},{1.2,1}}],(*draw    cotangent    line,i.e.,line
tangent at (0,1)*)
Line[{{Cos[rads π],0},{Cos[rads π],Sin[rads π]}}],(*draw    sine
line*)
Line[{{0,Sin[rads π]},{Cos[rads π],Sin[rads π]}}],(*draw    cosine
line*)
txt["Sine",{Cos[rads π]+tol,Sin[rads π]/2},{0,0},{0,-1}],
txt["Cosine",{Cos[rads π]/2,Sin[rads π]+tol}]
},Axes->True,PlotRange->{{-.3,1.2},{-.3,1.4}},Ticks->None]
```

The last line of the code specifies two additional options. It specifies the range to plot. It also turns off the display of tickmarks on the axes.

The notebook that corresponds to this chapter includes an animated version of the graphic illustrating these four trigonometric functions. The code above goes inside the command **Animate**[commands, {variable, minimum, maximum}]. The variable that **Animate[ ]** uses is **rads**, with a minimum of zero and maximum of two. The result is that the point on the circle is spinning and the points on the tangent and cotangent lines appear and follow the movement of the point on the circle. The code inside the **Animate[ ]** command has some minor changes from the above code. It includes **If[ ]** commands to display the points on the tangent and cotangent lines when the angle is at the opposite side of the circle. The code does not display the labels. Finally, the code does not have a limited range to display and does not turn off tickmarks.

This subchapter refreshed your recollection of the basic trigonometric functions, which graphs use quite frequently. You also learned how to use graphics primitives to display an elaborate graphic and how to define a function and then use it.

# 2 Derivatives and Optimization

Newton and Leibnitz independently and at about the same time discovered the calculus, a new field of mathematics on about 1680. This has been a breakthrough for mathematical analysis. The calculus analyzes the changes of functions as their variables undergo infinitesimally small changes.

One of the crucial concepts of the calculus is the derivative, the slope of the function at a particular location. Mathematica calculates derivatives with the command D[equation, variable]. The equation can have symbolic form rather than numerical. The command that produces the second derivative of a function is D[equation, {variable, 2}]. Higher order derivatives follow the same pattern.

Quantitative social science authors use derivatives to find the optima of the settings they model in mathematical terms. If the analysis indicates a set of costs and benefits from an activity, for example, the resulting equation may mean that individuals follow the conduct that maximizes that equation. The equation's maximum is at the point where its first derivative is zero and its second derivative is negative. Vice versa, if the analysis seeks the minimum, that appears where again the first derivative equals zero but the second derivative is positive.

This relation between the derivative and the extremes of functions follows from the geometry of slopes and peaks. If a function has a maximum, the slope of the function is positive (upward) before the peak (the maximum). In other words, before the peak, the derivative of the function (its slope) has positive values. After the peak, the function points down and has a negative slope, hence a negative derivative. The derivative crosses from positive values to negative values exactly at the maximum, the point where the function stops increasing, is about to start decreasing, and, momentarily, has slope of zero. Therefore, the derivative slopes down. The slope of the derivative, which is the second derivative of the original function, is negative.

Naturally, seeing an illustration makes this much clearer. Suppose that care costs $(c + c^{1/2})/2 - 1$ and the expected cost of accidents is $1/c$. At very low values of care, the cost of accidents is enormous. At very high values of care, the cost of accidents is trivial. What is the optimal care?

In Mathematica, we can (1) form the equation and store it in a variable, (2) take its derivative with D[ ], (3) solve it with **Solve[ ]**, (4) take the second derivative, and (5) test its value at the solution. The corresponding code is:

```
eq1 = (c+c^{1/2})/2 -1+1/c
d1eq1 = D[eq1, c]
```

```
s1 = N@Solve[d1eq1 == 0, c]
d2eq1 = D[eq1, {c, 2}]
N[d2eq1 /. s1]
```

Notice that the function **N[ ]** asks for a numerical value rather than the precise solution, which involves a long sequence of fractions and square roots. Also, a new abbreviation appears in the last line, slash-period. The output of **Solve[ ]** is a rule, **c->1.1...** rather than a number. The slash-period asks Mathematica to apply that rule. The result is that Mathematica substitutes **c** with the value that **Solve[ ]** produced. Therefore, when evaluating the second derivative, Mathematica produces its value at the extreme.

These commands reveal that an extremity of the function is at about 1.7. The value of the second derivative is positive at that point is about 1.15. Therefore, this extremity of the function is a minimum.

A figure, however, illustrates all this much more clearly. The corresponding code and the figure it produces follow. The code uses the **Plot[ ]** command, using the syntax **Plot**[list of equations, {variable, minimum, maximum}, *options*]. The options are numerous. (1) The option **AspectRatio** sets the ratio of the height to the width of the graph. When it is set to **Automatic** the units on the axes have the same size (rather than scaling the vertical axis to produce a graph that has the golden ratio as its aspect ratio). (2) The option **PlotRange** specifies which region of the graph appears. (3) The option **PlotStyle** has a list with three elements that specify how to draw the three lines that correspond to the three plotted functions. The first function only received the command **Black**. The second and the third receive a list of two specifications, that they should also be black and that they should follow two different dashing patterns. (4) The option **ImageSize** specifies the size of the graph. (5) The option **BaseStyle** receives a list specifying font and size for the elements that **Plot[ ]** produces, which in this instance applies to the labels of the tickmarks. (6) The **Prolog** option allows additional graphics primitives in the plot. The **txt[ ]** function defined previously puts labels on the three functions. Finally, (7) inside the **txt[ ]** function, the format that Mathematica uses for functions, **f[c]**, gets the **TraditionalForm** command. The result is that those functions take the conventional attributes of mathematical formatting. The square brackets become parentheses, the letters become italicized, and the apostrophes become primes.

A rudimentary display of the graph would not necessarily need to specify any options. The result is much better, however, after the graph receives this additional care (Fig. 4).

```
fntandsz = {FontFamily->"Roman", FontSize->.7 ftsz};
dp1 = Plot[{eq1, d1eq1, d2eq1}, {c, 0, 3},
AspectRatio -> Automatic, (*axes use equal units*)
PlotRange -> {-1, 2},
PlotStyle -> {Black, (*how to draw the three functions*)
{Black, Dashing[{.03, .01}]},
{Black, Dashing[{.03, .01, .002, .01}]}},
ImageSize -> 4.5 imgsz,
BaseStyle -> fntandsz, (*use the above specification of fonts
and sizes*)
Prolog -> {(*placing the text using Prolog*)
```

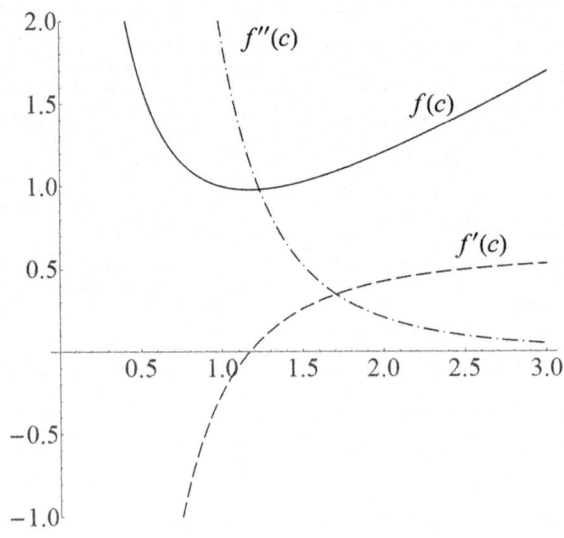

**Fig. 4** Function, first, and second derivative

```
txt[TraditionalForm@f[c],{2.3,1.5}],
txt[TraditionalForm@f'[c],{2.6,.65}],
txt[TraditionalForm@f''[c],{1.3,1.9}]
}]
```

This subchapter reminded you how to use derivatives for solving optimization problems. You saw sample code for getting first and second derivatives, for solving the first and checking the value of the second, and for plotting them.

# 3    Differential Equations and Transitions in Time

The power of the calculus to calculate the slope of functions can also be deployed in the opposite direction, to obtain functions that follow given specifications about their slopes. This subchapter of the calculus is *differential equations*. The primary application of differential equations is in obtaining processes of fading transitions. An example of such transitions in law comes from changes in legal regimes about affirmative action.[1] A simpler example, however, illustrates.

Suppose some impediments hinder one group of a population from obtaining a certain occupation. A new policy or technology removes all the impediments. How will the disadvantaged group's participation in this occupation change if no policy accelerating it exists?

The path to a solution may begin by making an assumption about the participation of the disadvantaged group. For example, assume that when the impediment was present, the pool of employees had a certain composition, with the disadvantaged group participating at the rate of, say, $w_0 = 1\%$. After the removal of the impediments, employers hire proportionately from the two groups. Call the disadvantaged group's fraction of the population $d$. The unknown participation of the

---

[1] *See, e.g.*, Nicholas Georgakopoulos & Colin Read, *Transitions in Affirmative Action*, 19 INT'L REV. L. ECON. 23 (1999).

disadvantaged in this workforce is a function of time, $p(t)$, which we do not know.

After the impediments disappear, the employers hire the disadvantaged with the rate $d$. However, an assumption is necessary for when employers hire. One approach is to ignore growth of both the population and employment and assume that employers only hire to replace retiring employees. Then, the model needs the ratio of the populations and the rate of retirements. Suppose the disadvantaged population is $d = 40\%$ of the population and the annual quit rate is $q = 10\%$ of the workers who retire every year. Then, we know that 10% of the workers will be replaced by a group that is 40% disadvantaged. Therefore, the pool of the employed will change by $qd$. However, as the fraction of disadvantaged increases in the pool of the employed, some of the retirements will be retirements of the disadvantaged. Thus, the net rate of change of the composition of the workforce is the ratio hired $q$ times the ratio minority $d$ minus the ratio retiring $q$ times its ratio minority, $p(t)$. Notice that this only tells us about the rate of change of this workforce, nothing about its actual composition. In other words, from the derivative of the workforce we seek the workforce itself. Essentially, we know that the derivative of participation is

$$p'(t) = qd - qp(t).$$

Once we have this derivative, we have stated the problem as one of differential equations and Mathematica can solve it. Then we can graph it. The command for solving differential equations is **DSolve** [list of equations, function sought, variable that changes]. The list of equations contains two equations, the assignment of the initial state to $p(0)$ and the above formulation of its derivative, making the corresponding code:

```
DSolve[{p[0] == w0, p'[t] == q d - q p[t]}, p[t], t]
```

The solution contains the Euler constant, symbolized with a double-struck e. In Mathematica, the Euler constant is a capital **E**; hitting escape-e-escape produces the lowercase double-struck e, which

Mathematica also recognizes. The Euler constant is also known as the base of the natural logarithm. The solution of the differential equation is:

$$p(t) = e^{-qt}\left(de^{qt} - d + w_0\right).$$

Despite the repeated appearance of the unusual Euler constant, plotting this is simple. We need, however, to give values to the variables $d$, $q$, and $w_0$. We do this with a list of rules. The list follows a slash-dot after the function of workforce participation and substitutes values for the quit rate, set to 10%; the initial participation in the workforce of the disadvantaged group, set to zero; and the disadvantaged fraction of the population, set at 40%:

```
Plot[E^(-q t)(-d+d E^(q t)+w0)/.{q->.1,w0->0,d->.4}, {t,0,40}]
```

The resulting plot is Fig. 5. The period of time plotted spans from zero to forty. The minority's participation in the workforce approaches its fraction in society but that approach keeps becoming slower. In the language of mathematics, the participation is asymptotic to the disadvantaged group's fraction of the population, per the example 40%. The group's workforce participation comes ever so close to 40%, but never quite becomes 40%.

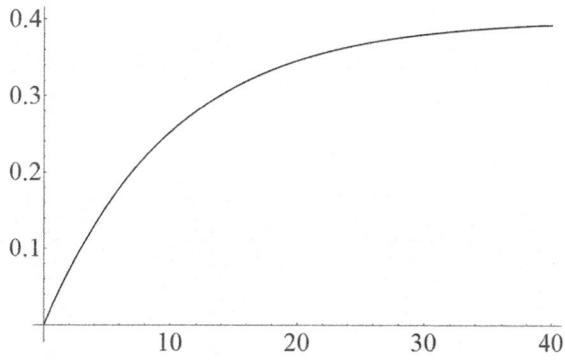

**Fig. 5**   Workforce participation over time

A richer analysis could add layers of complexity. A temporary policy of affirmative action, i.e., preferential hiring from the disadvantaged group, could exist over some period of time. Also, the population from which employers hire could be seen as finite, with its composition changing in reaction to hiring policies.

# 4    Conclusion

This chapter reviewed graphically three massive topics in mathematics: trigonometry, derivatives, and differential equations. Trigonometry is central to producing graphics because it allows the calculation of coordinates from angles. Derivatives are important for the social sciences because they allow the calculation of optima, the minimum or maximum values of equations capturing social activity. Differential equations enable the researcher to derive the equations that govern processes about which we only know their rates of change.

# 5    Exercises

1. Display a triangle with

```
Graphics[{
  Polygon[{
    {0.0},{1,0},{.5,.86}
  }]
}]
```

It turns out that Mathematica has a shortcut for rotating coordinates. The mechanism of this rotation uses matrix algebra, which this book avoids. Nevertheless, deploying the method does not require us to understand matrix algebra. Read the help file on the command **RotationMatrix** and notice the use of the period command (".") which corresponds to a matrix algebra operation. The order in the sequence of the rotation matrix and the coordinates are important. Use **RotationMatrix** to add a copy of this triangle in the graph so the copy is rotated $\pi/4$ counterclockwise. Place the rotated triangle after the original triangle in the list of graphics primitives and separate them with a color specification and a specification of partial opacity (for example, **Yellow, Opacity[.6]**).

2. Place your prior code inside a **Manipulate[ ]** or **Animate[ ]** command. Replace the angle of rotation of the second triangle with a variable (perhaps named **angle**), the range of which the last term of the **Manipulate[ ]** command defines. The result is that you can adjust the rotation using a slider.

What should be the range of the values of the angle of rotation?

Unless you specify that Mathematica should draw blank space in a graph, when it produces a graph, it only displays the relevant space. Thus, Mathematica produces a smaller region when the triangles of the graph overlap and a greater region when they are in opposition. The result is not elegant as you move the slider for the angle in the **Manipulate[ ]** command. How can you avoid having the graph change size as the angle of rotation changes? (Hint: consider **PlotRange**).

# 4

# Money and Time

Our first step into finance examines the movement of money through time, also called *time-value of money* and *discounting*. The analysis begins by considering that banks and credit markets always offer an alternative for the placement of funds. Recognizing this means that a payment on any past or future date can be equivalent to an investment or disinvestment today, where the terms of the credit markets apply for the difference in time.

## 1    Periodic Compounding

The usual terms of loans from banks and deposits at them specify that interest becomes part of principal once each month. The subsequent month, the calculation of interest will use the larger amount. The conversion of interest into principal is called *compounding*.

A simple example explains compounding. Consider that a bank offers interest of 12% per year with monthly compounding. A deposit of $100 in this bank will grow to more than $112 by the end of a year. At the end of the first month, the deposit will have earned interest of

© The Author(s) 2018                                                            **43**
N. L. Georgakopoulos, *Illustrating Finance Policy with* Mathematica,
Quantitative Perspectives on Behavioral Economics and Finance,
https://doi.org/10.1007/978-3-319-95372-4_4

one-twelfth of the year, or 1%. Applied to the deposit, which is $100, the interest earned is $1. Compounding makes the interest into principal so that, for the second month, the calculation of interest will use a deposit of $101 rather than $100. The result is that the second month's interest is a little more than $1, and so on.

Take a step back to calculate the general equation that gives the interest at the end of one month. Call the principal deposited at the present time $p$, the annual interest rate $r$, and the number of compounding periods in the year $n$. At the end of one month, saying that the deposit has earned one-twelfth of the year's interest rate means $pr/n$. To calculate the value of the account we add the principal, getting $p + pr/n$. Mathematicians consider this inelegant, take $p$ as the common factor, and write $p(1 + r/n)$. Call that $f_1$.

For the second month, interest accrues on this entire amount. So at the end of the second month, the account has $f_1(1 + r/n)$. Substituting $f_1$ from above gives $p(1 + r/n)^2$. Each month that passes increases the exponent by one. Because we measure time in years, however, we need to transform years into months by multiplying by the number of months in the year. The result is the formula for the future amount $f$ into which the present value $p$ grows under an interest rate $r$ compounded $n$ times per year in $t$ years:

$$f = p\left(1 + \frac{r}{n}\right)^{tn}. \tag{1}$$

The next step is to reverse time. What is the present value that we must deposit into such an account in order to find a future amount $f$? We need to solve the above equation for $p$. Divide both sides by the parenthesis raised to its exponent. It cancels on the side of the $p$ and gives:

$$p = \frac{f}{\left(1 + \frac{r}{n}\right)^{tn}} = f\left(1 + \frac{r}{n}\right)^{-tn}. \tag{2}$$

The second version of this formula may not be as familiar. Raising the parenthesis to a negative number is equivalent to dividing by it raised to the positive number. While this book avoids this type of notation, in more quantitative texts it appears often.

Having obtained both the formula for moving value forward through time and backward, periodic compounding holds no more mysteries. The nature of time, however, is another matter.

## 2 Continuous Compounding

I hope we agree that periodic compounding seems to have a hint of an artificial flavor. The division of the year into twelve months is arbitrary. Arbitrary is also the very notion of counting time in years. Yet, the notion of earned interest becoming principal seems natural.

This takes us back to the mathematical concept of limits. Let the number of periods go toward infinity in the future value formula (1).[1]

The result is $pe^{rt}$.

In other words, if compounding occurs an infinite number of times each year, amounts grow to their product with the Euler constant raised to the product of the interest rate and time. As long as time and the interest rate use the same unit of measurement, such as the year, the result is correct. To put it in conventional mathematical notation, the future amount $f$ under a rate $r$ as the number of compounding periods goes to infinity is:

$$f = \lim_{n \to \infty} p\left(1 + \frac{r}{n}\right)^{tn} = pe^{rt}. \tag{3}$$

Solving for $p$ gives the formula for moving time backward in time, as we did above to derive (2):

$$p = \frac{f}{e^{rt}} = fe^{-rt}. \tag{4}$$

These four formulas govern the movement of value through time. Recent versions of Mathematica have built-in functions for dealing with discounting problems: The help files for **TimeValue[ ]**,

---

[1]In Mathematica:

```
limit[p(1+r/n)^tn,n → Infinity]
```

`Cashflow[ ]`, and `Annuity[ ]` are well worth studying for those who will use these functions repeatedly.

# 3    Annuities

An interesting time-value of money question asks the value of an annual fixed payment, an annuity. One path to the answer would be to discount every payment to the present. However, a shortcut is to ask what deposit in a hypothetical bank would produce the same payments. If the payments continue to infinity—making the annuity into a perpetuity—this produces the correct answer.

A bank paying interest rate $r$ will pay to the depositor the deposit $p$ times $r$ each year. The annual payment $a$ is equal to $pr$, and solving for the deposit by dividing both sides by $r$ gives the value of the annuity: $p = a/r$.

Consider the value of an annuity that grows at some rate $g$. The present value of the first payment is $a/(1 + r)$. The present value of the second payment is $a(1 + g)/(1 + r)^2$, of the third $a(1 + g)^2/(1 + r)^3$, and so on, with the exponents increasing by one each year, to infinity. Reducing the infinite sum to a simple formula using Mathematica is very easy. The difficulty lies in stating the sum. Since $(1 + g)^0$ equals one, we can include this term in the first numerator with the exponent starting from zero. The sum becomes

$$\sum_{i=0}^{\infty} \frac{a(1 + g)^i}{(1 + r)^{i+1}}. \tag{5}$$

The corresponding Mathematica command is straightforward.[2] The sum gets restated as $a/(r - g)$. Very importantly, however, notice that the denominator becomes zero and negative when $g$, growth, reaches

---

[2]The Mathematica command is:

```
Sum[ a(1+g)^i / (1+r)^(1+i) ,{i,0,Infinity}]
```

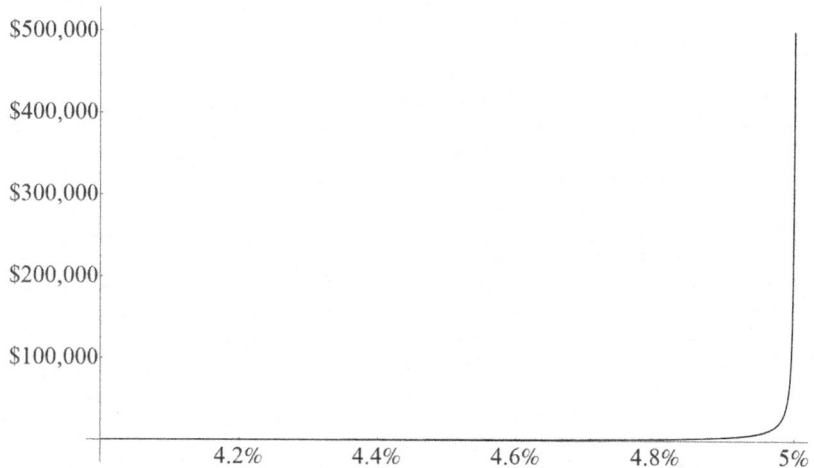

**Fig. 1** The annuity fallacy

and passes the value of $r$, the interest rate, making those values mean-ingless. This is a solution for a special case, only true if the growth is significantly smaller than the interest rate. If the growth is faster, we will need to discount actual future payments to the present during the fast growth. If the growth subsides, the remaining annuity follows the sim-ple valuation. If the growth may continue, then one must resort to a different valuation of the remaining value, likely based on the capital asset pricing model, which is the object of Chapter 5.

To observe the meaninglessness of this formula for the value of an annuity, merely as growth $g$ approaches the interest rate $r$, observe the value of an annual payment of $5 when the interest rate is 5% and the growth rate goes from 4% to almost 5%. Figure 1 is the result, a line fairly flat at about $100 that explodes to pass $500,000 as growth passes 4.9% and approaches 5%.

Consider first the setting with no growth. What deposit in a 5% environment produces $5 per year? Obviously, that is a deposit of $100. It makes some sense that any growth increases the value of the annuity above that. As growth approaches 5%, however, the value of the annuity essentially tends toward infinity, which only makes sense as a mathemat-ical abstraction.

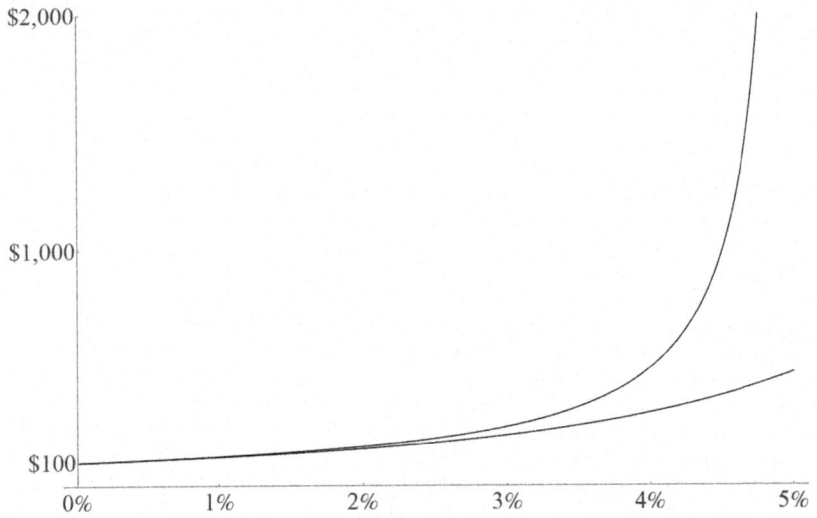

**Fig. 2**  The value of an infinite growing annuity and a hundred-year one

A return to realism is available by calculating the present value of this annuity if it were finite but lasted many years, say one hundred years. Essentially, the only change to Eq. (5) is that instead of the sum going to infinity it extends to one hundred. This is mathematically unappealing because it results in 100 terms, in stark contrast to the elegant simplification of the infinite annuity. Figure 2 compares the value of the hundred-year annuity to the infinite one with growth rates from zero to 5%. The vertical axis is limited to a maximum value of $2,000. Obviously, the lower curve corresponds to the finite annuity. If instead of going to infinity the annuity lasts a hundred years, the present value of those payments does not exceed $400, whereas we saw in the prior figure that when the growth rate approaches the interest rate, the infinite annuity reaches absurdly large values.

The difference between the two curves of Fig. 2 is the present value of the payments beyond the first one hundred years. The value of those payments is what drives the spike in the valuation of the infinite annuity. Infinity may be a fascinating mathematical topic, but some real applications of infinity can be quite misleading.

# 4    Internal Rate of Return

Investors naturally would tend to ask an apparently simple question: I do not know what the correct discount rate may be, just let me know the rate that this investment offers. In finance, the rate that the cashflows of the project itself offer is the *internal rate of return*. For mathematical finance, however, this is a frustrating question because it is mathematically unappealing. Returning to the analogy to a hypothetical bank, consider a project's cashflows, investments into the project as negative amounts (deposits into the hypothetical account) and earnings from the project as positive amounts (withdrawals from the account). The internal rate of return is the interest rate that causes the hypothetical account that reflects investments and earnings to have a zero balance at the end.

Consider a simple example, the investment of $100 today for receiving $10 each year for two years, then in three years receiving $110. The corresponding cashflows are −100, 10, 10, 110. It should be obvious that this corresponds to a 10% rate under annual compounding because it is equivalent to placing $100 in a hypothetical account, removing the interest of $10 each year for two years, and then on the third year closing the account and finding $110, the initial deposit of $100 and the last year's interest of $10.

The corresponding equation for the present value of those payments (cash flows) is not simple:

$$-100 + \frac{10}{1+r} + \frac{10}{(1+r)^2} + \frac{110}{(1+r)^3}.$$

Equating to zero and solving it for $r$ in Mathematica gives three solutions, $.1$, $(-3 - i\sqrt{3})/2$, and $(-3 + i\sqrt{3})/2$. The boldface $i$ is the imaginary unit, i.e., the square root of minus one. Thus, the last two solutions are imaginary numbers and do not apply, but Mathematica provides them as valid mathematical solutions. If instead of annual compounding we use periodic compounding or continuous compounding, the results become correspondingly more complex.

The command **FindRoot [ ]** gives only the real solutions, avoiding the mess of the imaginary ones.[3] Exercise 4 asks you to create a function that would handle IRR questions in a more convenient way.

## 5    Risk

The formulas about the time-value of money convey a false sense of security. I hope that was especially apparent in the discussion of the infinite annuity with fast growth. A crucial component of the value of future payments is their certainty. Yet, nowhere in these equations do we see an attempt to capture risk or uncertainty. The problem is that risk is a very difficult concept.

Many treatises say that discounting assumes no risk or low risk, which makes some sense. In the real world, however, settings with low or no risk are few and could change into higher risk settings. Most finance scholars, for example, treat promises of future payments by the US government as safe claims, which may be close enough to accurate for many purposes. However, we should not forget that in the late seventies and early eighties the United States had high inflation and that in the last twenty years Congress has tried to limit the ability of the United States to issue more debt to finance its operations.

Returning to the mathematics of finance, however, does give a way to handle risk, unlike the above worries about inflation and governmental disagreements, which have no answers. Modern finance handles risk with the Capital Asset Pricing Model, which is the object of the next chapter.

---

[3]The command for obtaining only the real solution from Mathematica is **FindRoot [** eqn, {variable, initialguess}] which, using a guess of 7%, becomes in the instance of annual compounding

$$\texttt{FindRoot}[-100 + \frac{10}{1+r} + \frac{10}{(1+r)^2} + \frac{110}{(1+r)^3} == 0, \{r, .07\}]$$

which may solve the present question but is somewhat clumsy.

# 6    Conclusion

This chapter studied how the value of money moves through time by considering functionally equivalent uses of simple lending or borrowing. The chapter derived the formulas for periodic compounding and continuous compounding, valued annuities that are fixed and ones that grow. Finally, the chapter derived the internal rate of return of a sequence of cash flows.

From the perspective of policy, the transition to modern methods of valuation initially had to overcome the inertia of adherence to precedent. Courts were fairly quick to accept the new methods.[4]

# 7    Exercises

1. Produce the figures using Mathematica.
2. Calculate the following using a function you define based on the equations of this chapter and using the built-in Mathematica commands **TimeValue[ ]** and its relatives (mostly **EffectiveInterest[ ]**):
   2.1  The present value of 914.09 under a monthly [compounded] 5% rate. (Throughout these exercises, the usage "monthly rate" means the "monthly-compounded annual rate" and similarly for other compounding periods, such as quarterly or daily. In other words, do not multiply by twelve to convert the monthly 5% rate into an annual 110% rate.)
   2.2  The annual rate that is equivalent to a quarterly rate of 5%.
   2.3  The quarterly rate that is equivalent to an annual rate of 5%.

---

[4]*See, e.g.*, Weinberger v. UOP, Inc., 457 A.2d 701 (Del. 1983) (where the Delaware Supreme Court accepts discounting and flexible methods of valuation based on modern finance rather than insisting on the "Delaware block" method of its precedent).

2.4  The continuous rate that is equivalent to a monthly rate of 5%.

2.5  The annual rate that is equivalent to a continuous rate of 5%.

2.6  The annual rate that causes an investment to double in ten years.

2.7  The amount of time that causes an investment to double under an annual rate of 7.2%.

2.8  The monthly rate that causes an investment to triple in ten years.

2.9  The continuous rate that causes an investment to double in ten years.

2.10  The amount of time in which an investment doubles using a continuous rate of 10%.

2.11  What is the relation of the last two answers to the natural logarithm of two? (This is an observation, not a present value question.)

3. Ford Corporation issued a bond with the following terms: Ford will pay back the principal of $1,000 to the holder of one unit of the bond on July 16, 2031. Twice annually, Ford will pay the holder half of the annual interest, calculated using the interest rate of 7.45%. On January 16, 2018, this bond is trading at 128% of its par value, i.e., at $1,280. These exercises are probably easiest to solve using a spreadsheet program, provided that it has the functionality of Excel's Goal Seek command (Under Data/What If Analysis/Goal Seek).

3.1  What actual return would someone who buys this bond on January 16, 2018, and holds it till it matures enjoy? State this as a periodic compounding rate. What compounding period seems to make sense?

3.2  Using the bond's return, as you calculated it in the prior exercise, calculate the present value of each payment under this bond. Weighing the date of each payment by the present value of the amount to be paid on that date, calculate the weighted average date of repayment. The time to that date is called the bond's *duration*.

3.3 Repeat the above two steps assuming continuous compounding. To obtain the continuous rate you will need to add a spreadsheet cell summing the present values of all the cash flows—i.e., the negative initial investment and the sequence of positive amounts that the holder will receive—and then use Goal Seek to set that cell equal to zero by changing the rate. (Note that Excel's macro language allows you to write a macro that would run when any of the cells in a range changes, so you could automate this task.)

3.4 Repeat the above—i.e., calculate the periodic rate, the continuous rate, and the duration under each—for the actual date and price of this bond when you are answering this exercise.

3.5 Calculate the periodic and continuous rates and the durations of Ford's 6.625% October 1, 2028, bond and of Ford's 4.75% January 15, 2043, bond, according to their prices when you answer this exercise. (Simpler version: Assume the time is January 16, 2018, and they are trading, respectively, at 118% and 100% of their par value.)

3.6 Discuss the phenomena you observe.

3.7 If you were Ford's Chief Financial Officer, and Ford had some extra cash and some bonds of each of these three (either unissued or that Ford had repurchased), would you use the cash to buy back some of these bonds? Which ones? If Ford needed additional funds, which bond would you sell?

4. Write a function that takes as its arguments (a) a list of dates and payments and (b) the number of compounding periods (e.g., 1, 12, infinity) and returns the internal rate of return.

The main difficulty lies in forming the sum that **FindRoot[ ]** would solve. You could use Mathematica's built-in **Cashflow[ ]** function. Consider the following two examples from the documentation page for the command **Cashflow**:

```
TimeValue[Cashflow[{{2,    100},    {5,    200},
{7.5, 200}, {10, 500}}], .09, 10]
FindRoot[TimeValue[Cashflow[{-1000,    100,
200, 300, 400, 500}], r, 0] == 0, {r, .05}]
```

The first takes as its first term a list of times (2, 5, 7.5, …) and payments at those times (100, 200, 200, …), the interest rate of 9% and the final time (10). It returns the future value of those payments at time 10.

The second has a list of six cashflows and calculates their internal rate of return.

To choose the appropriate compounding period, also read on the function **EffectiveInterest[ ]**.

5. Use Mathematica's help files: What is the difference between the Mathematica functions **Annuity[ ]** and **AnnuityDue[ ]**?

# 5

# The Capital Asset Pricing Model

The prior chapter's tools of moving value through time do not address risk. In a crude sense, more risk suggests a greater rate of discount or interest. Under greater risk a future amount is worth today less than if it were safer. Conversely, a loan to a riskier debtor requires a greater interest rate to compensate for the probability that the debtor will fail and be unable to pay the debtor's obligations. Until the development of the Capital Asset Pricing Model, these adjustments for risk were not scientific. The valuation of risky assets, paradigmatically stocks, was a mystery.

The Capital Asset Pricing Model is an extraordinary intellectual achievement of towering proportions. The complexity of the CAPM (pronounced cap-EM) requires a gradual approach, in four steps.

The first step toward the Capital Asset Pricing Model is to recognize the benefits and limitations of diversification. In terms of the history of ideas, probability theory concludes that diversification reduces risk. However, diversification in stocks does not eliminate risk whereas diversification in truly independent random events can reduce risk to any arbitrarily low level.

© The Author(s) 2018
N. L. Georgakopoulos, *Illustrating Finance Policy with* Mathematica, Quantitative Perspectives on Behavioral Economics and Finance, https://doi.org/10.1007/978-3-319-95372-4_5

The second step accepts that even diversified investors bear some risk and seeks to understand what risk that is. The risk of each stock divides into diversifiable idiosyncratic risk and into non-diversifiable systematic risk.

The third step applies the statistical method of linear regression to the relation of each stock to the stock market. The result is that we can calculate how sensitive to the stock market each stock is. In other words, a measure exists that tells investors who consider adding a stock to their portfolio, how much of the risk of the entire system that stock brings.

Finally, the CAPM combines all these steps and recognizes that diversified portfolios with the same degree of systematic risk must offer the same returns. The measurement of systematic risk and its pricing become related. The CAPM uses market data to infer the relation of expected returns and systematic risk for every asset, resolving the problem of valuing risky assets.

# 1    Diversification and Its Limits

The benefits of diversification are palpable in an example of coin tosses. Having a purse of $500, and simple double-or-nothing bets on coin tosses, consider the question of into how many bets you should split the purse. Putting the entire $500 on a single bet, a single coin toss, gives two potential outcomes, $1,000 in the case of a win and $0 in the case of a loss. Each has probability of 50%. Dividing the purse into several bets (diversifying among coin tosses) quickly reduces the probability of extreme outcomes. Dividing the purse into four bets means that four wins or four losses are necessary to reach the extremes of $1,000 or $0. Half the time you win the first bet. Half of that half of the time you also win the second bet, and so on. Winning all four bets has probability of one-sixteenth, half raised to the fourth power, or .0625.

The reduction of the probability at the extremes comes with the creation of additional possible outcomes and additional probability weight to the outcomes near the average. In the example of the four bets, several paths of wins and losses lead to an outcome of $250 or $750, which correspond to a single win or a single loss. More paths produce two wins to lead to an outcome of $500. Rather than counting paths, we can use the binomial distribution to find the probability of winning one, two, or three tosses. We seek a function of the distribution that gives the probability of one of its possible outcomes. This is the probability distribution function. The answers come from the probability distribution function of the binomial distribution for a number of tosses $t$, each with probability .5 of winning.[1]

In other words, define a function which takes two inputs, $x$ and $t$. The function equals the probability distribution function of the binomial distribution for $t$ trials (tosses) with a .5 probability of winning each evaluated for $x$ wins. This function gives the probability of winning any specific number $x$ of bets when the total number of bets into which the purse was divided is $t$. Asking what is the probability of winning one toss when we divided our purse into four bets means setting $x$ equal to one and $t$ equal to four.

Besides the probability of each outcome, we must also be able to calculate the payout of each outcome. That is the number of wins, $x$, times 1,000 divided by the number of tosses.[2]

We can then see the effect of the change from (a) letting the entire purse ride on a single toss to (b) splitting it into four bets.

The vertical axis of Fig. 1 holds the probability of each outcome. The horizontal axis holds the final payout, from winning nothing to winning every toss and receiving $1,000. The points along each curve represent each possible outcome. If we place the entire purse on a single toss, the curve marked "1 toss," then the possible outcomes are two, winning nothing or winning a thousand. Each has probability .5. The line marked "4 tosses" tracks the probabilities and payouts in the case where we divide the purse into four bets on four different tosses. Then the possible outcomes are five, from zero to a thousand in increments of $250. Their probability ranges from one-sixteenth at the extremes to almost .4 at the center.

---

[1]In Mathematica this can be:
**prob[x_, t_]: = PDF[BinomialDistribution[t, .5], x].**

[2]In Mathematica this can be:
**pay[x_, t_]: = x 1000/t.**

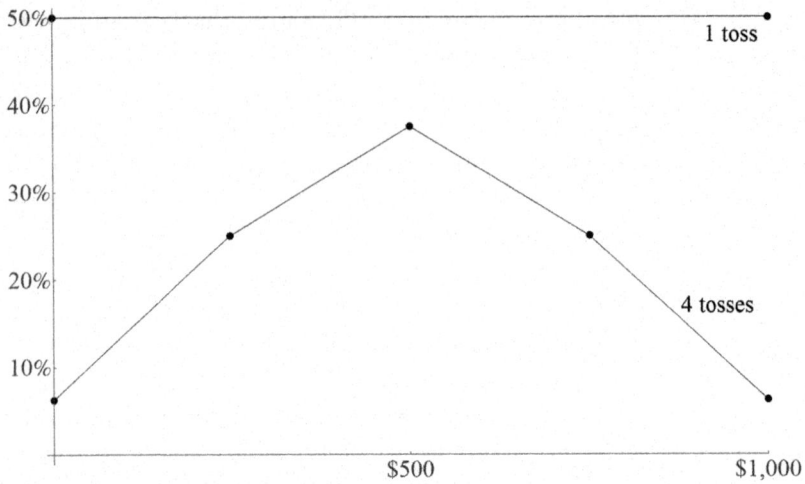

**Fig. 1**  Betting $500 on one or four coin tosses

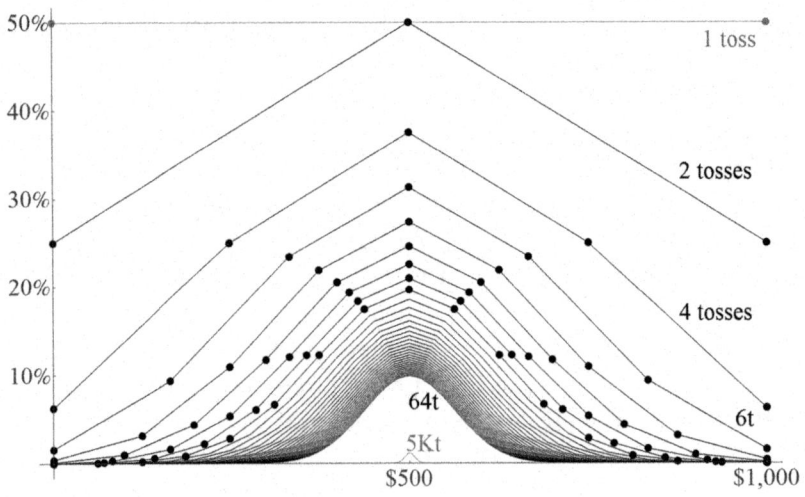

**Fig. 2**  Probabilities and payouts as tosses increase

Figure 2 continues this process, creating a curve with the probabilities and payouts for dividing the purse to a number of tosses equal to every even number from two to sixty-four. The figure marks the points on the curves only up to sixteen tosses. The figure also shows, grayed-out, the

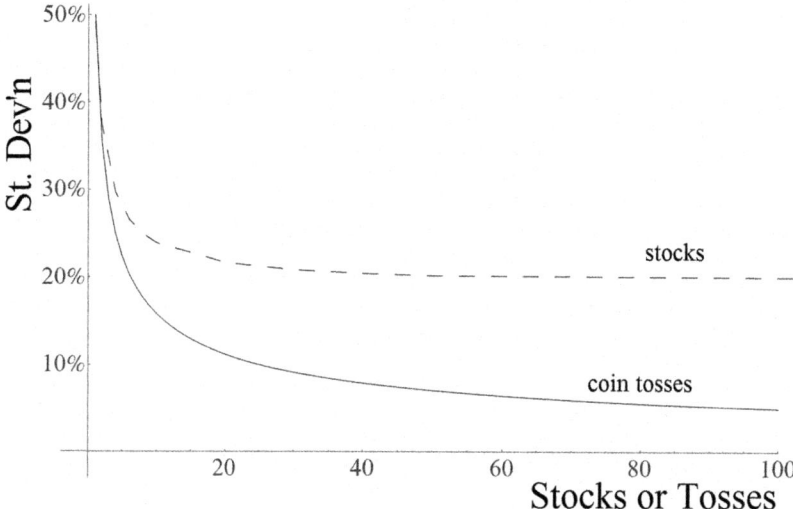

**Fig. 3**   Diversification in stocks and coin tosses

single toss case, and, to illustrate that risk keeps declining, the case of five thousand tosses, for ten cents each, marked 5Kt. As the number of tosses increases, the extremes become extraordinarily unlikely. As the purse is split into more tosses, the extremely unlikely outcomes spread from the edges toward the center. The probability weight gets concentrated around $500. The shape of the line that connects the probabilities approaches the familiar bell curve of the normal distribution. By diversifying, the risk of the bet changes dramatically.

The conventional way to measure the risk for the bettor as diversification increases is the standard deviation of the final payout. On the basis of the above, we should expect as the number of coin tosses increases, the standard deviation of the payout to decrease.

Researchers have turned to test the equivalent proposition on stocks.[3] How much does diversification reduce the standard deviation in stocks (for annual returns)? Figure 3 compares the reduction of standard deviation in coin tosses and stocks as we diversify from one to a hundred

[3]E. J. Elton & M. J. Gruber, *Risk Reduction and Portfolio Size: An Analytic Solution*, 50 J. Bus. 415 (1977).

sources of risk, either tosses or stocks. The two curves already diverge at five sources of risk. Diversification among coin tosses can bring standard deviation close to zero. Diversification among stocks has a floor at about 19%. Regardless how diversified investors' portfolios may be, investors cannot reduce their risk below an annual standard deviation of 19%.

The advantages of diversification are crucial, despite its limitations. Even among stocks, diversification reduces risk, as measured by standard deviation, from over 50% to about 19%, which is enormously valuable.

Nevertheless, the limitation is important. The fact that diversification in stocks cannot break this floor at about 19% poses the question of the next step toward the CAPM: Why cannot diversification in stocks eliminate nearly all risk, as diversification does in coin tosses?

## 2    Undiversifiable Risk

Diversification cannot eliminate risk in stocks because all stocks, as stakes in the underlying businesses, depend on a common source of risk, the overall performance of the economy. Granted, each stock carries significant risk that is particular to that business. Damage to its productive assets, being overtaken by competitors' innovations, personal difficulties of its key personnel, etc., are risks that influence that particular business. The financial literature calls them idiosyncratic or firm-specific risks.

The dependence of stocks on the common source of risk that is the overall economic performance makes them unlike pure gambles. Gambles such as the coin tosses of our example, have the feature of being *independent*, in probability theory language. Such gambles are independent because their outcomes do not depend in any way on outcomes of other uncertainty, the other tosses. They have no relation to each other or to any common source of uncertainty. By contrast, stocks are correlated and are not independent because their performance depends on the economy's overall performance. When the economy booms and spending is forthcoming, all firms tend to do well. When the economy is in a recession and most cut spending, all firms tend to fare poorly.

The undiversifiable 19% risk comes from the common uncertainty of the performance of the entire economy. Firms cannot avoid the consequences of booms and recessions. Some firms may be somewhat protected and may mitigate the downturns, but firms cannot avoid recessions completely. The downside of recessions also has the upside of booming times, when the rising tide raises all boats. Financial literature calls this common undiversifiable risk *systematic* or *market risk*.

The inability of investors to diversify market risk seems like a dead end. However, statistical methods shed light on it.

# 3   Measuring Market Risk

Having realized that stocks carry undiversifiable risk, because all business activity depends on overall economic activity, the next question is whether all stocks are equally sensitive to the economy. If some stocks are more sensitive to the economy and some are less sensitive, then investors could select among them and only diversify among the less sensitive ones, for example.

Statisticians have developed a mathematical technique for calculating the line that passes closest to a set of points on the Cartesian plane. While the matrix algebra that this technique uses is complex, effectively all numbers-oriented computer programs have it built-in as a function, making it easy to deploy.[4]

When trying to apply this idea to the relation of a stock to the economy, data on the stock is plentiful due to its daily trading, but comparable data on the economy seems nonexistent. Nevertheless, a measurement does exist which financial theory and practice has accepted as an appropriate substitute, the index of the performance of the entire stock market. Out of the several such indices, theory and practice have converged on the Standard and Poors 500 index, an index composed of the 500 stocks with the largest market capitalization on

---

[4]In Mathematica, the corresponding command is **LinearModelFit[ ]**. In spreadsheets it is = linest( ).

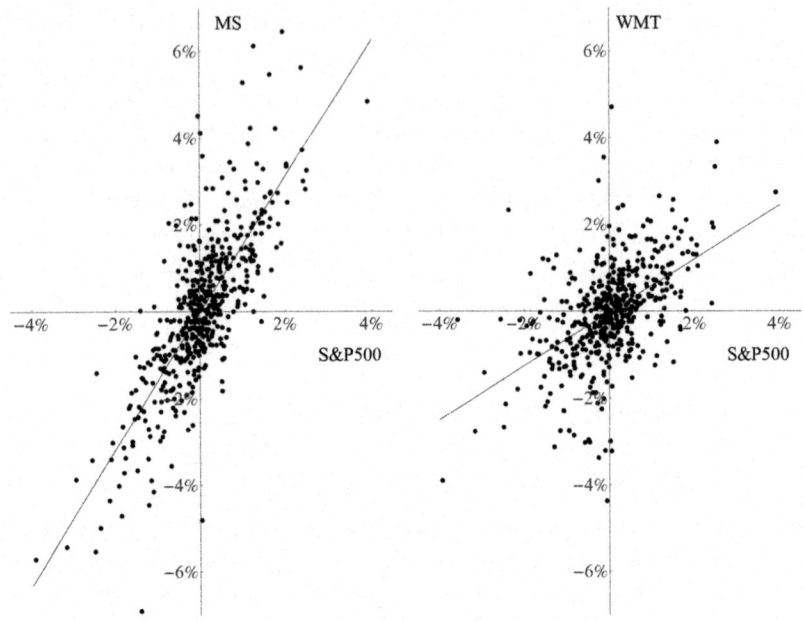

**Fig. 4**   Morgan Stanley (MS) and Walmart (WMT) against the index

the New York Stock Exchange and the NASDAQ, in other words, the 500 highest-valued listed US businesses.

One can gather the daily changes of the price of a stock and of the S&P500 index, consider each pair a point on the Cartesian plane, and draw the line coming closest to all those points. Figure 4 does that for two stocks, Morgan Stanley and Walmart.[5] Notice that they have

---

[5]For example, the corresponding Mathematica code to obtain the MS data and convert it into daily percentage changes is:

```
ms2yrs = FinancialData["MS", {DateObject[DatePlus[
    DateObject[{2016, 10, 22}], -2365]],
    DateObject[DateObject[{2016, 10, 22}]]}]];
ms2ydd = Table[{
    ms2yrs[[i + 1, 1]], (*keep the date coordinate*)
    ms2yrs[[i + 1, 2]]/ms2yrs[[i, 2]] -
    1}(*today's price as a fraction of yesterday's-1*)
    , {i, Length@ms2yrs - 1}]
```

different sensitivity to the index. The slope of the line captures the general tendency of each stock to move with the market. If one line had a slope of a half and the other a slope of two, we could say that for a given change in the index, we would expect one stock to change half as much as the index and the other twice as much as the index.

Statistics uses Greek letters as parameters for the formula of the line that passes closest to a set of points. The line on the plane defined by the horizontal $x$-axis and the vertical $y$-axis is given by $y = \alpha + \beta x$. When applied to stocks and the index, the index occupies the horizontal axis and the stock the vertical axis. The stock's sensitivity to the index is the beta of the equation. In stark contrast to the financial community's practice of devising colorful names for all sorts of phenomena, from black swans and poison pills to white knights and cash cows, the collective imagination seems to have been stumped by the notion of sensitivity to the index.[6] The parameter, beta, became the universal name for sensitivity to the index.

Figure 4 shows the daily changes in the stock against the changes of the index for Morgan Stanley (symbol MS, on the left panel) and

---

After placing the pairs of daily percentage changes of the stock and the index in a list called **mstosp**, the command for obtaining the line is

```
msols = LinearModelFit[mstosp, x, x]
```

Then, the code that produces the graph can be

```
Show@Graphics[{ Point@mstosp, (*points*)
   Line[{
      {-.04, Normal[msols] /. x -> -.04}, (*linestart*)
      {.04, Normal[msols] /. x -> .04}}] (*lineend*)
   }, Axes -> True,
AspectRatio -> Automatic]
```

Invoking the **Normal[ ]** of the linear model produces the equation and the code that follows it, slash dot $x$ to a value, values the equation of the line at that value. Can you determine whether the data account for dividends?

[6]Black swans are rare large changes in prices that have destabilizing consequences; *see, e.g.,* Nassim N. Taleb, The Black Swan: The Impact of the Highly Improbable (2d ed. 2010). Poison pills are a security that functions as a defense against a hostile takeover of the corporation; *see, e.g.,* Werner L. Frank, Corporate War: Poison Pills and Golden Parachutes (2011). White knights are friendly acquirers, used to stave off hostile acquirers; *see, e.g., id.* Cash cows are firms that produce stable and predictable profits; *see, e.g.,* The Boston Consulting Group on Strategy: Classic Concepts and New Perspectives (Carl W. Stern & Michael S. Deimler, eds., 2006).

Walmart (symbol WMT, on the right panel). In each, the vertical axis holds the changes in that stock and the horizontal axis holds the changes in the index. Each point corresponds to one day's pair of changes of the index and the stock. Each graph captures two years of trading, 503 trading days and points, from October 23, 2014 to October 21, 2016. Each graph also plots the line that fits best the data according to the statistical method of linear regression. The slope of that line is the beta of each stock. The formula that defines this line for Morgan Stanley is about $0 + 1.6x$. For Walmart the formula is $0 + .6x$. Morgan Stanley has a beta of 1.6, and Walmart has a beta of 0.6. A holder of a diversified portfolio with betas about that of Morgan Stanley would experience about 160% of the fluctuations that the overall market experiences. A holder of a diversified portfolio with betas like that of Walmart would experience 60% of the fluctuations of the overall market.

The measurement of the sensitivity of each stock to the market leads directly to the ability to handle risk in a diversified portfolio.

# 4    The Capital Asset Pricing Model

To pull all the above together in a single pricing model, the Capital Asset Pricing Model relies on a key realization that has three components. First, the measurement of beta means that investors can construct diversified portfolios that have specific betas. Second, investors can also construct portfolios with specific betas by mixing the market portfolio and lending or borrowing, especially lending to the US government by buying US treasuries. Third, an actual and a synthetic portfolio with the same beta should be equivalents. Therefore, the expected return of the market and of US treasuries determine the expected returns for every beta. The Capital Asset Pricing Model calculates those returns.

For example, consider that we construct a diversified portfolio with a beta of a half. Consider also that we can synthesize a portfolio with a beta of half by putting half the funds in treasuries and the other half in an index fund. The real and the synthetic portfolio should behave very similarly and should have the same return. Because we know the expected return of the market portfolio and of treasuries, we can calculate the expected return of the beta half portfolio. Using this method, we can calculate the expected return for any asset given its beta. That is the Capital Asset Pricing Model.

The equivalence of a synthetic and actual portfolio flows from the function of beta as a measure of sensitivity to changes of the market. Saying that a beta half portfolio will change half as much as the market is as much a tautology as saying that a beta half portfolio has half the risk of the market. To synthesize a portfolio of any beta, an investor must hold a fraction beta of the portfolio's funds in the index (likely an index mutual fund) and the remaining as riskless treasuries. For betas greater than one, it appears that the investor must borrow and again invest a beta fraction of the portfolio's funds in an index fund. Because borrowing would be at higher rates than those of the treasuries, an adjustment may have been necessary. However, even for betas greater than one, the equivalence holds using the rate of the treasuries because the portfolio with the greater beta, mixed with treasuries, can synthesize a beta of one. Thus, the orthodox CAPM applies the rate of the treasuries throughout. The evidence suggests that using a slightly higher rate is more realistic.

Since the synthetic portfolio must have the same return as the actual portfolio with an arbitrary beta, the question becomes simply establishing the expected return of the synthetic portfolio. That is beta times the expected return of the market plus one minus beta times the interest rate of the treasuries, $R_f$. The customary appearance of the CAPM, however, is a transformation of that formula. It splits the return of the market, $R_M$, into the premium of the market return over the interest rate of the treasuries, $R_M - R_f$. The result is the formula that gives the expected return, $R_\beta$, for stocks with a known beta as the risk-free rate plus beta times the market premium:

$$R_\beta = R_f + \beta(R_M - R_f).$$

The most powerful empirical validation of the Capital Asset Pricing Model comes from a statistical study that grouped stocks into ten

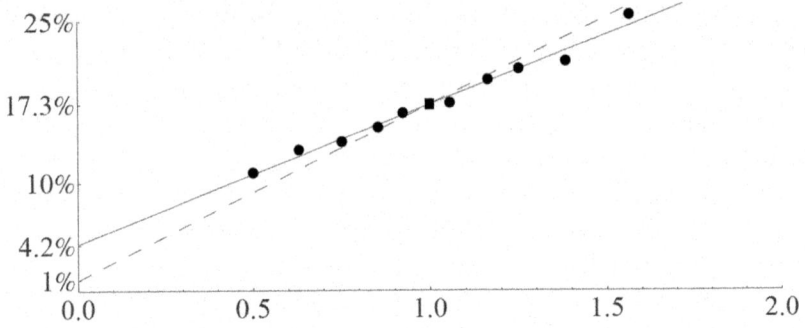

**Fig. 5**  Empirical test of the CAPM with 1931–1965 data

portfolios, by beta decile.[7] Studying returns from 1931 to 1965, the results make Fig. 5. The horizontal axis holds beta and the vertical axis annual returns.[8] The ten points are the ten portfolios, and the square marker is the market. The solid line is the "capital market line" that they define, which produces a higher return for the zero beta asset than the risk-free rate. The risk-free rate over that period was negligible; the figure simulates it as 1% to draw a dashing line that would correspond to the orthodox CAPM.

Not only do the returns of the portfolios align by beta but also so does their risk. Although the financial literature does not display standard deviations and neither did the original paper, adding error bars to these points corresponding to the standard deviations of their evidence, shows that risk aligns almost perfectly. Figure 6 adds error bars spanning one standard deviation below and above each portfolio's return and that of the market (the square). The figure also fits two lines to each end of the error bars and displays those lighter and dashing.

Finance scholars felt uncomfortable that the evidence did not fit perfectly the CAPM. The capital market line was too flat. However, the model's reliance on the risk-free interest rate is more questionable itself.

---

[7]Fisher Black, Michael C. Jensen & Myron Scholes, *The Capital Asset Pricing Model: Some Empirical Tests, in* STUDIES IN THE THEORY OF CAPITAL MARKETS (Michael C. Jensen, ed., Praeger Publishers Inc., 1972).

[8]The original paper used monthly returns, here multiplied by 12 rather than compounded by raising to the 12th power. Compounding the returns would have made the graph qualitatively different than the original figure.

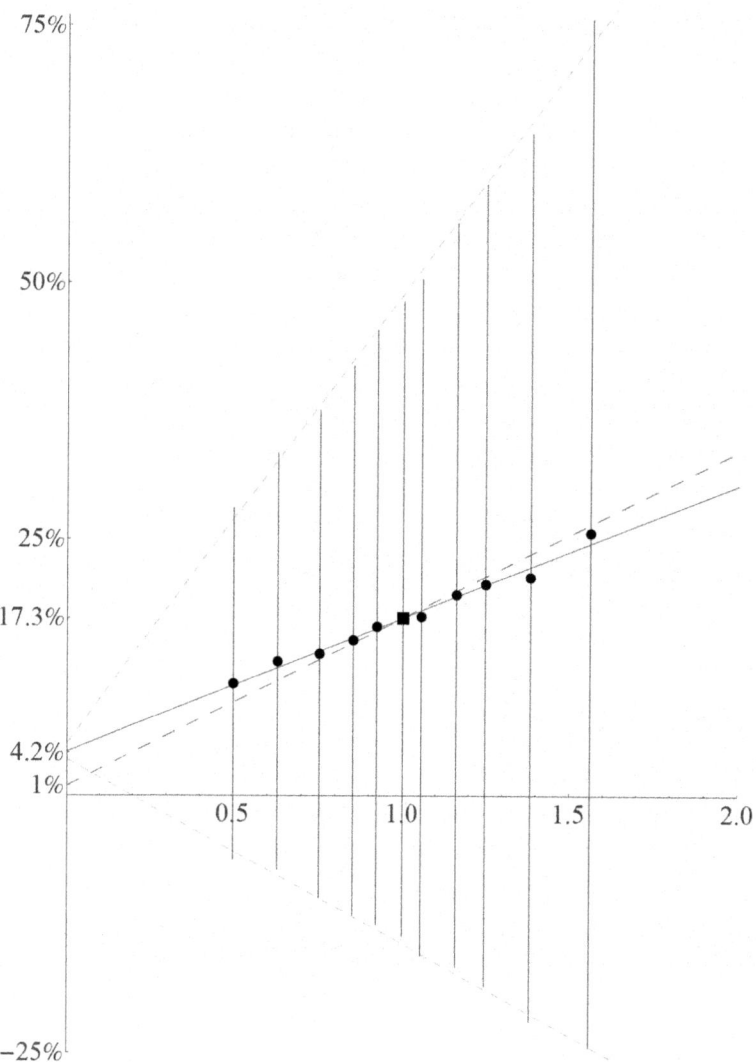

**Fig. 6**  The CAPM evidence with standard deviations as error bars

If investors in practice consider that riskless lending occurs at a different, higher rate, for example, that of a diversified portfolio of high-
grade corporate bonds, then reason suggests that this higher rate would
ground the pricing of assets.

Moreover, investors cannot borrow at the interest rate of the treasuries. Some of the arbitrage transactions that ensure the accuracy of the CAPM, rely on borrowing. This again argues that a higher rate would drive pricing. The discrepancy between the evidence and the model should not surprise. Rather than seeing the glass as half empty, the data reveals it as mostly full. Especially realizing how abstract the CAPM is, the degree of agreement between the model and the evidence is striking.

The compelling nature of the CAPM becomes vivid through a trading opportunity that arises if we imagine a mispricing. The mispricing will give rise to an arbitrage opportunity that ensures pricing in accordance with the CAPM. Suppose that a set of stocks, enough to form a diversified portfolio, with some beta, say 1.2, are cheaper than indicated by the CAPM. This means that this diversified portfolio will outperform the synthetic portfolio with the same beta. Effectively, every trader has an incentive to sell 120% (the beta fraction) of the fraction of their market portfolio that they devote to such trades and buy 100% of that fraction of the mispriced portfolio. After some time, when the mispricing dissipates, a virtually riskless gain will accrue for the trader. The evidence from 1930 to 1965, however, is even more compelling. The CAPM operated despite that such trades could not happen—the CAPM was developed in the sixties as was the computing power required for it.

Despite what seems as resounding validation, finance scholars have continued to comb the evidence for models that perform more accurately than the CAPM. Various metrics—such as the ratio of the market price to book value,[9] the debt-to-equity ratio,[10] and the price-to-earnings ratio[11]—have been argued to produce models that explain returns better than the simple CAPM.[12] However, some of

[9] *See, e.g.*, Donald B. Keim, *Size-Related Anomalies and Stock Return Seasonality*, 12 J. Fin. Econ. 13 (1983).

[10] This line of literature begins with Laxmi Chand Bhandari, *Debt/Equity Ratio and Expected Common Stock Returns: Empirical Evidence*, 43 J. Fin. 507 (1988).

[11] This line of literature begins with S. Basu, *Investment Performance of Common Stocks in Relation to Their Price-Earnings Ratios: A Test of the Efficient Market Hypothesis*, 32 J. Fin. 663 (1977).

[12] *See also* James Ming Chen, Econophysics and Capital Asset Pricing: Splitting the Atom of Systematic Risk (2017).

these metrics likely identify errors that the market may tend to make rather than having a principled economic explanation for their effect. Indeed, some of these effects have later disappeared, like the size effect. Others, like the debt-to-equity ratio, may be due to the effect of concerns that the CAPM ignores, such as taxation or executive incentives. The higher valuation of more leveraged firms may correspond to their smaller tax exposure, to the disciplining effect of debt on management, or to the (not entirely unrelated) reduced likelihood of a takeover. Granted, in an ideal world, firms would adjust their borrowing to reach the optimal leverage. However, some frictions in the real world should not cause alarm and do not invalidate the well-reasoned and amply supported CAPM.

# 5    Reconciling with the Practice of Capitalization Using the P/E Ratio

Whereas scholarly finance justifiably focuses on the CAPM, practice continues to use the cruder yardstick of pricing by the capitalization method, multiplying earnings by the Price-to-Earnings ratio of comparable firms. One estimates next year's earnings for the business to be valued, finds a few comparable firms with listed stocks, calculates their average ratio of their price to their earnings for next year, and multiplies the business's earnings by that ratio to obtain its value according to the capitalization method.

Take the example that we estimate our business to have earnings of $500,000 next year. The comparable firms have an average P/E ratio of 12.5. The value of the business according to the capitalization method is 12.5 times $500,000 or $6,250,000.

However, the price-to-earnings ratio of different stocks has enormous variation. As a result, the capitalization method is very sensitive to which firms one considers comparable. As we should expect based on discounting principles, the variation of P/E ratios corresponds mostly to expectations of future growth. Thus, for firms to be comparable (for using their P/E ratio to value a business), those firms should have similar growth expectations. Yet, if the focus is on estimating future growth, then the simplicity of using P/E ratios for valuation, rather than discounting, mostly disappears. The shortcut is only convenient within industries provided that most firms in an industry have similar expectations of growth. However, if one uses the capitalization on a young or small firm, then that firm is likely to have better (but also more risky) growth expectations than those of the more seasoned and larger firms that likely are the industry comparables.

Despite the imperfections of the capitalization method, and perhaps due to its persistence in practice, finance scholars try to improve it. For example, one line of thought explicitly refines the analysis by including the expected growth rate.[13] Another approach uses expected growth rates to better select comparable firms.[14] Generally, the capitalization method can seem accidentally satisfactory and surprisingly resilient in practice.

# 6     Conclusion

This chapter explained the Capital Asset Pricing Model. The foundation was diversification in independent gambles. Stocks, however, did not let diversification have as complete an effect in reducing risk. The limited effect of diversification is due to the joint dependence of all businesses on the performance of the economy. The statistical method of linear regression, moreover, can measure the corresponding undiversifiable risk. The measurement of that risk, beta, allows us to think of portfolios that expose their holders to different amounts of undiversifiable risk. The CAPM flows from this realization. Each asset's beta determines its return as the risk-free return plus beta times the market premium, i.e., the market's return minus the risk-free return.

---

[13]*See, e.g.*, Peter D. Easton, *PE Ratios, PEG Ratios, and Estimating the Implied Expected Rate of Return on Equity Capital*, 79 ACCT. REV. 73 (2004).

[14]*See, e.g.*, Sanjeev Bhoral & Charles M.C. Lee, *Who is My Peer? A Valuation-Based Approach to the Selection of Comparable Firms*, 40 J. ACCT. RES'CH 407 (2002).

# 7    Exercises

1. Using one of the several sources of historical data or Mathematica's financial data (in Excel, copy and paste the data, which is limiting; in Mathematica the command for obtaining the data is **FinancialData**[*stocksymbol,* {*startdate, enddate*}]), download a few years of historical data for two stocks of your choice and the S&P500 index (often under the symbol SPX), and:

   1.1 Calculate the daily, weekly, monthly, and annual changes of the stock and the index (while preserving their dates).

   1.2 Calculate their average and standard deviation.

   1.3 Calculate the implied standard deviation of longer terms by multiplying by the square root of time (e.g., the implied weekly standard deviation is the daily one multiplied by the square root of five), i.e., calculate implied weekly, monthly, and annual from the daily; implied monthly and annual from the weekly; and implied annual from the monthly. How do the actual standard deviations compare to the implied ones?

   1.4 Match by date the changes of the stock to those of the index and calculate the beta using daily returns, weekly, monthly, and annual ones. Compare those to the beta reported by the various financial news sites. Discuss the differences.

   1.5 Produce a graph akin to that of Fig. 4.

2. Suppose that the risk-free rate is 3% and the expected return of the market is 9.8%. What is the expected return of an asset with a beta of .8?

3. A diversified portfolio that has existed for ten years has experienced an average annual return of 14%. The risk-free return has averaged 2% and the expected stock market return has been 10%.

   3.1 For what beta of this portfolio are its returns consistent with the Capital Asset Pricing model?

   3.2 The manager of this portfolio claims to be an excellent stock picker. How do you evaluate this claim if the beta of the portfolio has been 1 and how do you evaluate it if the portfolio's beta has been 1.7?

4. Table 1 holds the results of the Black, Jensen, & Scholes study of stock returns from 1935 to 1965 (behind Figs. 5 and 6). For each decile of stocks, the table gives beta, monthly return, and monthly standard deviation. For the entire market, the corresponding values are 1, .0142, and .0891. Recreate Figs. 5 and 6. Keep in mind that standard deviation increases with the square root of time, so it would be wrong to multiply the monthly values by twelve. Rather, multiply them by the square root of twelve.

**Table 1**  Beta decile returns 1935–1965

| Decile beta | Monthly return | Monthly st.dev. |
| --- | --- | --- |
| .4992 | .0091 | .0495 |
| .6291 | .0109 | .0586 |
| .7534 | .0115 | .0685 |
| .8531 | .0126 | .0772 |
| .9229 | .0137 | .0836 |
| 1.0572 | .0145 | .0950 |
| 1.1625 | .0163 | .1045 |
| 1.2483 | .0171 | .1126 |
| 1.3838 | .0177 | .1248 |
| 1.5614 | .0213 | .1445 |

5. With a large team of classmates, repeat the study that produced the above table and Figs. 5 and 6 for a perhaps shorter period that reaches to a more recent time. Perhaps first devise a filter that reasonably and unbiasedly reduces the number of stocks to a more manageable figure. Allocate deciles and sub-periods among the team for filtering, sorting deciles, and processing. For example, you may choose the period from 2000 to 2015, use each prior year starting from 1999 to calculate betas and sort stocks into deciles, then calculate results for the next year, repeat till you reach the final year, then aggregate the results for all the years. Coordinate early and often, ensuring that your analyses are compatible and can be combined.

# 6

# Options

## 1    Types of Options

One of the great contributions of mathematical finance is the call
option valuation formula, also known as Black–Scholes or Black–
Scholes–Merton. It was the object of the 1997 Nobel Memorial Prize in
Economics. This chapter approaches financial options, the substance of
the call option valuation formula, and gives its graphical representation.
Then, recognizing that corporate securities can be analogized to financial options, the chapter returns to corporate settings, including the idea
that reorganizations could be done through the distribution of packages
of options.

Real options (i.e., non-financial ones) are familiar and ubiquitous.
I have some free time this evening, therefore I have numerous real
options: to go to a restaurant, to a show, to catch up on some work.
Contractual real options have more structure. A contract with a garage
may give me the option to park there at some times during the day.
Options can be seen as antithetical to obligations. In the above examples, one is not obligated to go to a restaurant nor to deliver a car for
parking at the garage. However, the other side in a contractual option

© The Author(s) 2018
N. L. Georgakopoulos, *Illustrating Finance Policy with* Mathematica,
Quantitative Perspectives on Behavioral Economics and Finance,
https://doi.org/10.1007/978-3-319-95372-4_6

does have an obligation. The garage has undertaken the obligation to be open and take the car for parking (although the garage may have stipulated some exceptions to this obligation, perhaps depending on capacity and holidays).

Financial options are even more structured. Financial options (this chapter will use *options* for financial options hereafter) rest on an *underlying* financial asset subject to frequent trading on an exchange, such as a stock. Options have an *expiration date*, by when their owner must make the decision to *exercise* them, and a *strike price*, the price at which the owner gets to buy or sell the underlying asset.

Options are rights to buy or sell the underlying asset by the expiration date at the strike price. *Call* options are rights to buy the underlying asset at the strike price by the expiration date (mnemonic: to call someone and have the asset delivered at the strike price; like calling for a pizza). *Put* options are rights to sell the underlying asset at the strike price by the expiration day (mnemonic: to put the asset in someone's hands and demand payment; if you sell a put on a pizza, make sure you have hot pads!).[1]

The other sides, the sellers or *writers* of options, have contingent obligations, depending on whether the buyers choose to exercise their options. The writer of the call has the contingent obligation to sell the underlying asset at the strike price by the expiration date. The writer of the put has the contingent obligation to buy the underlying asset at the strike price by the expiration date. Unlike stocks and bonds, the corporate issuer is rarely involved in most traded options. Other traders take the two sides in the options market.

## 2    Payouts

The financial position of the sides of call and put options, thus, depends on the price of the underlying asset at expiration. If the asset is priced more favorably than the right that the option gives—if the holder of a call

---

[1]Some options do not allow their holder to exercise them early. Those are called *European options*. The text describes the options that allow early exercise and are called *American options*.

can buy the asset for less than the strike price or the holder of a put can sell it for more—then the holder will not exercise the option. The option has no value. Otherwise, the option places its holder in a better financial position than without it. The holder is wealthier by the amount of the difference between the price of the asset on expiration and the strike price. The holder of a call is wealthier by the amount that the price exceeds the strike price. The holder of a put is wealthier by the amount that the price is below the strike price. For example, the holder of a $10 call that expires today on stock XYZ, which is trading at $16, is wealthier by $6 due to owning the call. The holder of the call can exercise the option, buy a share of XYZ stock spending $10, and sell it for $16. Vice versa, the holder of a $20 put on the same stock is wealthier by $4. The holder of the put can buy the stock for $16 and exercise the put option to sell it for $20. Because on expiration the value of the option becomes deterministic in this way, option contracts can settle by default by charging the writers the corresponding amount without requiring the holders to exercise them.

This clear financial effect of options on expiration leads to a graphical representation of the position of their holders and their writers. Figure 1 displays the four corresponding graphs, the financial position of the holder of a call and a put and the financial position of a writer of a call and of a put on the vertical axis as a function of the price of the stock at expiration on the horizontal axis. The interesting feature of all these graphs is the angle each has. At the price ranges where the options are not exercised, the payoffs are zero. On the other side of the strike price, the payoff line becomes diagonal, increasing for the holders and decreasing for the writers. (As an exercise, consider a graph of the round-trip effect of an option that will not be exercised till expiration and was sold for a given amount; how would this change the graphs of Figure 1?)

The angular shape of option payoffs makes them tools for designing and simulating payoffs of arbitrary shapes. Given a payoff shape, usually one can use options to produce the same payoff. This is how option analysis applies to corporate decisions. The different claimholders of a corporation, the holders of debt with different seniorities, and the holders of preferred and common stock, can be envisioned as holding bundles of options that simulate their payoffs, as applied in Exercise 6. Before turning to applications of this approach, this chapter makes a visual study of the call option valuation formula.

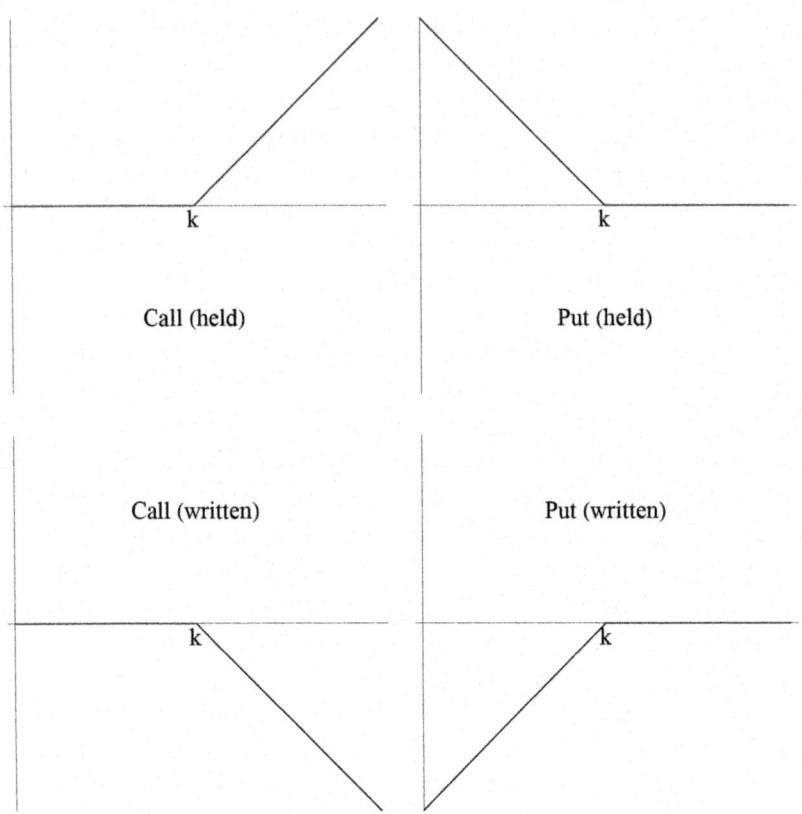

**Fig. 1**  Payouts to holders and writers of calls and puts

# 3    Call Option Valuation

The call option valuation formula is anchored on economic reasoning. Then the formula applies the mathematical power of probability distributions in combination with the call option payoff.

The fundamental premise is that it is possible to use a position in call options in combination with a position in the stock and lending or borrowing to produce a portfolio that has no risk. In the simplified example that comes, risk is eliminated. In more complex real-world settings, some risk might exist because, as prices change, hedgers might not be able to make some trades that would be necessary to keep risk at zero.[2] Because the portfolio has no risk, then the holder of this portfolio is not entitled to earn a return that includes a risk premium; the governing return is the risk-free rate. This is an incredibly important step because it avoids the need to estimate the stock's return, which means calculating the stock's beta and estimating the market return, which are all noisy signals and would, therefore, inject inaccuracy to the valuation of the call option.

To realize that a portfolio of calls and the underlying stock can become free of risk in a simple setting, take a stock that can have only two future prices, $8 and $12. The portfolio writes one call with a strike price of $8. How many shares must the portfolio own to be in the same position regardless of the final price? For each share that the portfolio owns (a) if the price that materializes is $8, then the stock will be worth $8 and the portfolio will owe zero because of the option (b) if the price that materializes is $12, then the stock will be worth $12 and the portfolio will owe $4 because of the option. Thus, we need to solve for $x$ the equation $8x - 0 = 12x - 4$ which reduces to $x = 1$. Suppose the expiration is immediate and the current price of the stock is $9. The option writer must spend $9 to buy one share of the stock, receives the price for writing the call, and the resulting portfolio has no risk and is worth $8 in either case. Therefore, writing the option must bring $1. If the option were trading for more, every options trader would rush to write calls, and if it were trading for less, none would. This first step of the analysis shows that the call can make a portfolio neutral toward risk and, at expiration, this dictates the price of the call.

At this point, the strength of this approach to pricing compared to one based on the probabilities of the payouts of the option is not yet visible. The $9 price of the stock reveals the market's estimate of the probabilities of each future price. The $9 price is only consistent with a 75% probability of an $8 price and a 25% probability of a $12 price

---

[2]For example, a hedger may need to trade if price changes by 1%. However, price can change by, say, 3% overnight while the markets are closed. Then, the hedger wakes up to a portfolio that is carrying some risk.

because .75 times 8 plus .25 times 12 is 9. This indicates that the call has a 25% chance of paying $4 and a 75% chance of paying zero and therefore should be worth $1. Thus, it appears that either approach to pricing is correct. Consider, however, the possibility of a price of the stock that makes no sense, outside the range from $8 to $12, say $13. Now probability is no guide but the arbitrage pricing is, giving a price of $5 for the call.

The second step is to add time until expiration. The risk-free portfolio can be created, but some time will pass before expiration and the realization of the stock's price on expiration. The portfolio invests $9 to buy one share of the stock, writes one call, and will wait, say, three months—one quarter—till expiration. Because the portfolio is neutral toward risk, the rate that the portfolio must earn is the risk-free rate. Therefore, we discount the $8 that the portfolio will be worth in three months using continuous compounding and the risk-free rate to find the value that the bundle should have today. The answer dictates the price of the call exactly as in the prior step. Thus, if the risk-free rate is 5%, the present value of $8 is that divided by the base of the natural logarithm raised to the amount of time (a quarter year) multiplied by the interest rate. The result is that the portfolio should have a value of $7.90 today. Therefore the value of the call is $1.10. The portfolio is worth $7.90 because it spends $9 to buy one share of the stock and receives $1.10 for writing one call.

To see the strength of the arbitrage-based option pricing, juxtapose an approach based on the expected value of the option itself. Suppose that the stock has a beta of 2, and the market return is 15%, making the market premium 10% (because the risk-free rate is 5%). Accordingly, the stock's $9 price implies that in three months the market according to the CAPM expects the stock to have grown by a continuously compounded annual rate of 25% (the risk-free rate plus beta times the market premium), namely to $9.58. The corresponding probabilities that this indicates for the stock being at $8 versus $12 are a probability of 60.49% for the stock being worth $8 and of 39.51% for being worth $12. Applying those to the call would falsely indicate that its pricing should be based on the idea that its expected value is $1.58 in three months (39.51% of $4). Option traders thinking that discounting that value by the rate indicated by the stock's beta would lead to the correct price for the call would be wrong. Even discounting $1.58 by the 25% rate indicated by the beta for a quarter indicates a present value of $1.48. Willing to pay this amount would, indeed, be too high by about 35%. The option traders who use the correct arbitrage-based pricing will sell as many calls to erring traders as the erring traders desire, without the accurate option traders bearing any risk.

Observe, instead, that the option price does correspond to the idea that the stock grows at the risk-free rate. Grown at 5% for a quarter would make the stock $9.11. The probabilities that this price indicates for the stock becoming worth $8 and $12 are 72.17% that the stock will be worth $8 and 27.83% that it will be worth $12, giving the call the expected future value of $1.113, which discounted at the risk-free rate makes it worth today $1.10, its correct value according to the arbitrage model. Despite that the market price of the stock indicates other probabilities govern the future prices of the stock (as the prior paragraph explained, 60.49% and 39.51%), the call pricing becomes correct by assuming that the stock will grow at the risk-free rate!

The transition from the simple example of two possible stock prices to the reality on an infinity of possible future stock prices can take two steps. The first step moves to a world of finite but several future prices. The second step transitions to infinite prices with no upper bound.

The first step is a simplified version of the binomial option pricing formula.[3] Somewhat simplified, it assumes that several cointosses will occur that will move the stock price up or down until the expiration date. Losing a toss pushes price down and winning pushes the price up and the probabilities need not be 50–50. The question becomes how to build the portfolio to be indifferent to the next cointoss, despite that many additional tosses will occur until expiration. The portfolio may need to be rebalanced after a cointoss in order to again become indifferent to the subsequent cointoss. Rather than develop this pricing, take

---

[3]In the Cox, Ross & Rubenstein binomial option pricing model the variability of the outcomes changes as well. *See* John C. Cox Stephen A. Ross, & Mark Rubinstein, *Option Pricing: A Simplified Approach*, 7 J. Fin. Econ. 229 (1979).

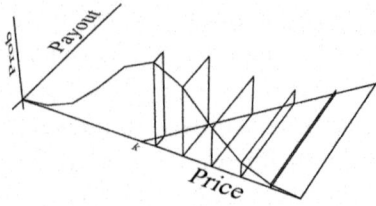

**Fig. 2**  A ten-cointoss binomial call

the shortcut of the prior paragraph, the assumption that the stock price will grow at the risk-free rate. Then, the objective is to establish the probabilities that make the expected terminal price equal to the amount to which today's stock price would become if growing at the risk-free rate (rather than that implied by its beta).

If, for example, ten cointosses will determine the final value of the asset, then after making the price grow at the risk-free rate, the simplified exercise will be to determine the present value of the future expected value of the call depending on the probability of each price materializing. The calculation of the future expected value of the call will be to multiply the payout by the probability of each price occurring. Figure 2 displays an example of a sequence of ten tosses, in six possible outcomes of which the call has value. The payout of the call in each of those cases is the depth dimension of the graph, labeled Payout. The probability of each outcome is the vertical dimension, labeled Prob[ability]. The future expected value of the call is the sum of the products of each probability of each price, which are the vertical dimensions of each rectangle, multiplied by the payout of the call on each price, which are the horizontal dimensions of the rectangles, in other words, the sum of the surfaces of the six rectangles. That sum, discounted to its present value, would be the value of the call (as opposed to the discounted expected value of the call with the stock growing at its true rate that accounts for the stock's beta). The strike price of the call is $k$. The floor of the figure is the payout to the holder of the call. The corresponding formula is

$$\frac{\sum_{x=k}^{c} \text{PDF(Binomial Distribution}(c, p), x)(x - k)}{e^{rt}},$$

where $c$ is the number of cointosses, and $p$ is the probability of success of each cointoss that makes the expected price be the one that corresponds to the stock growing at the risk-free rate. The possible future price of the stock is $x$. A way to read this formula is that it is the sum of the products of the probability of each outcome $x$ according to the probability density function of the binomial distribution for $c$ trials with probability of success $p$ multiplied by the difference $x$ minus $k$ with $x$ running from $k$ to $c$, divided by the base of the natural logarithm raised to the power of the product of the risk-free rate $r$ multiplied by the time till expiration $t$. The next step is to let the prices of the stock increase in number and not have a restriction on the maximum value they can reach. Keep in mind, however, that stocks do have a minimum value, zero; they do not take negative values due to limited liability.[4] While the normal distribution satisfies the two first conditions (it is continuous, thus offering an infinite number of possible outcomes, and it has no maximum, thus offering an unlimited upside) it fails the third because the normal distribution is also unlimited in its downside. The appropriate distribution is the lognormal, which is also continuous and has no upper bound but has zero as its lower bound. What is unattractive about the lognormal distribution is that it is not symmetrical, its mode does not coincide with its mean, and the mathematics required to handle it are very complex.

Figure 3 illustrates the lognormal distribution. The probability density functions (PDFs) of two lognormal distributions appear. The solid black one has a mean at $m = 100$ and standard deviation of $s = 10$. The dashing gray has the same mean but a larger standard deviation of 20. The solid black one would correspond to the safer stock and the dashing gray one to the riskier one. The range displayed is from 50 to 150 (as is in Figs. 4 and 5). If the PDF of a normal distribution, the

---

[4]Some scholars have proposed altering the regime of limited liability for corporations; *see, e.g.,* Henry Hansmann & Reinier Kraakman, *Toward Unlimited Shareholder Liability for Corporate Torts,* 100 YALE L.J. 1879 (1991); *but see* Nicholas L. Georgakopoulos, *Avoid Automatic Piercing: A Comment on Blumberg and Strasser,* 1 ACCT. ECON. AND L. (2011). https://doi.org/10.2202/2152-2820.1002. If limited liability were to be relaxed, the valuation of calls would need radical revision.

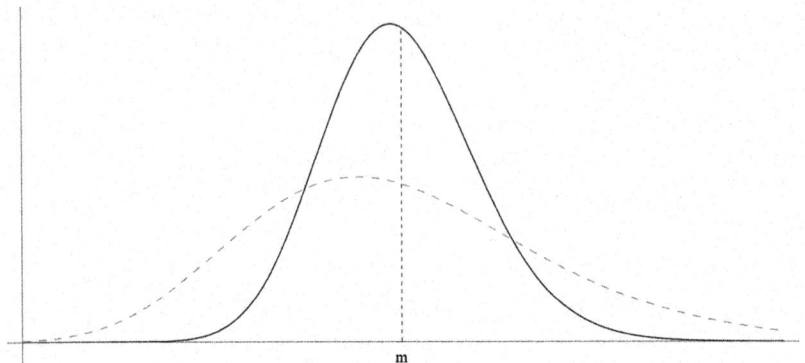

**Fig. 3** Comparison of two lognormal distributions

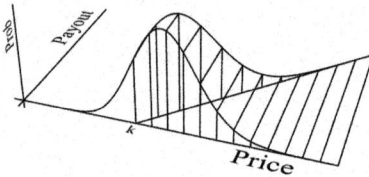

**Fig. 4** The solid of the call

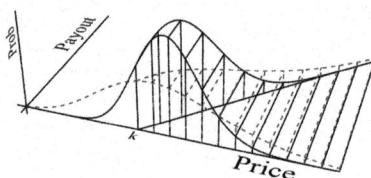

**Fig. 5** Two call solids

familiar bell curve, with the same mean were also superimposed, it would have its peak at the mean and be symmetrical.

Deploy the lognormal distribution in the same fashion as in the case of the binomial option. The mean of the distribution will not be where today's stock price would grow given its beta but where today's stock price would grow at the risk-free rate. Then, because the distribution is continuous, no finite number of rectangles will appear; rather than adding, we will use integration to calculate the volume of the wedge-like solid that an infinite number of rectangles would create. The standard deviation of the distribution will be that of the stock, which conventionally is given in an annual figure as a percentage of its price, and will be adjusted by the square root of time to the time remaining till the expiration of the option. The result is Fig. 4. The volume of the resulting wedge-like solid is the value that the call would be expected to have on expiration. That will be discounted to the present value to give the simplified estimate of the value of the call. The corresponding formula would be

$$\frac{\int_{x=k}^{\infty} \text{PDF}\left(\text{Lognormal}\left(m, s\sqrt{t}\right), x\right)(x - k)\mathrm{d}x}{e^{rt}} \tag{1}$$

where $m$ is the mean that makes the expected price be the one that corresponds to the stock growing at the risk-free rate, i.e., $m = \mathrm{p}e^{rt}$; and $s$ is the annual standard deviation of the stock. A way to read this formula is that it is the integral of the products of the probability of each outcome $x$ according to the probability density function of the lognormal distribution with mean $m$ and standard deviation $s\sqrt{t}$ multiplied by the difference $x$ minus $k$ with $x$ running from $k$ to infinity, divided by the base of the natural logarithm raised to the power of the product of the risk-free rate $r$ multiplied by the time till expiration $t$. If standard deviation is expressed as a percentage of price, as is the convention, then replace $s\sqrt{t}$ with $ps\sqrt{t}$. However, the lognormal distribution conventionally uses as its parameters not its own mean and standard deviation but that of the underlying normal distribution that is transformed by the natural logarithm into the lognormal. Determining the proper parameters that will produce a lognormal distribution with the desired mean and standard deviation is not straightforward. To avoid this problem, the Black–Scholes formula transforms $x$ and places it in the standard normal distribution from which the lognormal derives.

The derivation of the Black–Scholes formula does not follow the above simplification. Rather, the formula is the solution to a differential equation problem stating when the hedged portfolio has no risk. The Black–Scholes formula uses an auxiliary formula to help state the value of the call $c$:

$$c = p\text{CDF}(N(0, 1), \text{Aux}) - ke^{-rt}\,\text{CDF}\left(N(0, 1), \text{Aux} - s\sqrt{t}\right), \qquad (2)$$

where

$$\text{Aux} = \frac{\ln\frac{p}{s} + rt}{s\sqrt{t}} + \frac{s\sqrt{t}}{2},$$

where ln is the natural logarithm function, $p$ is the price of the stock, $k$ is the strike price, $s$ is the annual standard deviation as a fraction of the price, and $t$ is the time till expiration. Exercise 7 compares the value of the call according to the shortcut of the illustrations versus the Black–Scholes formula. The discrepancies are small.

One striking feature of option valuation is that more risk does not reduce the value of the call. Intuition from pricing ordinary securities would suggest that additional risk makes a security worth less. Rather, the value of calls becomes greater as risk increases because their insurance feature becomes more valuable. The graphical approach to call valuation also clarifies this intuition. As the stock becomes riskier, the solid may lose some volume at the peak if the strike price is below the mean as in these figures; but there the payout (depth) is limited. The additional risk, however, makes the tail of the distribution thicker. There, the payout (depth) is increasing, making this effect dominate. The calls on riskier stocks are more valuable. Figure 5 illustrates this effect. The solid black lines define the solid that corresponds to the call on a stock with an expected future price of 100 (if the stock grew at the risk-free rate) and standard deviation of 10%, the same call as the previous figure. The dashing gray lines define the solid of a call on a stock with the same expected price but greater risk, a 20% standard deviation. The strike price is $k = 90$ and the mean of the distribution is $m = 100$, two frames to the right. The range of stock prices displayed is from 50 to 150. The point of the figure is to show that calls on riskier stocks, all other equal, are more valuable. The two lognormal distributions of the stock prices are the same ones as in Fig. 3.

The final illustration on the pricing of calls focuses on the effect of time. If the stock price does not change, how does the value of the

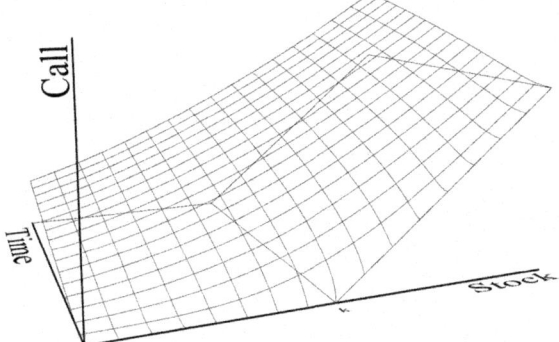

**Fig. 6**   The decay of the value of a call

call change as expiration approaches? How is this change different if the price is near the strike price ("at the money") compared to the strike price being far below ("in the money") or far above ("out of the money") the current price? Figure 6 illustrates the value of a call for different stock prices as the expiration time approaches. Time is in the depth dimension, labeled Time, with expiration being at the origin of the axes, near the viewer. The horizontal dimension, labeled Stock, is the price of the stock. The vertical dimension, labeled Call, is the value of the call according to the Black–Scholes formula. When expiration occurs, the surface becomes the payout graph for the call. Earlier, however, the value of the call is higher. Far from the strike price, the value of the call approaches the payout value in an almost linear way. This corresponds to an intuition that the insurance function of the call is small because the likelihood of a surprise large enough to change the pricing mode becomes vanishingly small. Near the strike price, however, the call retains much of its value until late and then drops fast. This corresponds to the intuition that the call protects its holder against the downside. The erosion of value of the call is akin to an insurance premium that its holder is paying with time.

   Figure 6 can also serve as the springboard for understanding the proper derivation of the Black–Scholes formula. Think about the slope of this surface, the value of the call, with respect to changes in the underlying stock price (as opposed to its slope with respect to changes in time, on which the prior paragraph focused). The perfectly hedged portfolio uses that slope when determining, for example, the number of shares to hold per written call. Then, the portfolio is indifferent to changes in the stock price because the changes in the value of the call cancel the effect of the changes in the stock. The Black–Scholes formula rests on determining that slope first. Then, solving the differential equation problem posed by that slope, produces the Black–Scholes formula.

Consider also how the slope of the value of the call with respect to stock price, changes over time. As time passes and the stock price changes, the ratio of calls written to stock held that hedges the portfolio's risk also changes. Thus, unlike the simplified case of the two prices with which we started, where the portfolio had no risk from inception to expiration, in the more complex setting of the real world, even a perfectly hedged portfolio will need adjustments as time passes as well as when the stock price changes.

An entire cottage industry of option pricing models followed the Black–Scholes formula, just as more complex formulations followed that of the Capital Asset Pricing Model of Chapter 5. The assumptions of the Black–Scholes formula are strict and imperfect. For example, the lognormal distribution has thinner tails than the markets' actual experience of the occasional large change, which matched the evidence that the Black–Scholes formula tended to undervalue far-out-of-the-money options. Some models addressed this by assuming that volatility was random ("stochastic"), others by adding occasional jumps into the price, and others by adding such features to the interest rate. A 1997 empirical comparison of several formulas argues that the most accurate is a formula addressing both jumps and stochastic volatility, with the Black–Scholes being the least accurate of the group.[5] The breakthrough of the Black–Scholes is that it established the method for pricing options, despite that it left room for additional sophistication.

---

[5] *See* Gurdip Bakshi, Charles Gao, & Zhiwu Chen, *Empirical Performance of Alternative Option Pricing Models*, 52 J. FIN. 2003 (1997).

# 4 Board Decisions in the Zone of Insolvency

The option approach reveals that corporate securities can also be seen as bundles of options on the underlying business or assets of the corporation (Exercise 6). The payout to holders of common stock looks like the payout to holders of calls. The payout to holders of debt looks like the payout to writers of puts. No surprise should come, then, from realizing that they have similar attitudes toward risk. An increase in the risk of the business can increase the value of the equity while reducing the value of the debt. Thus, the holders of the common stock of a business that is burdened with debt will enjoy an increase of the value of their holdings if the business takes more risk, whereas the holders of the debt will suffer a decline in value. The two groups of claimants have different preferences about how to manage the business.

This tension between the shareholders' desire to take more risk and the creditors' desire to take less risk reaches the courts usually as a challenge to decisions of the board of directors. The directors, elected by the shareholders and seeking to maximize shareholder value, accept some risk. The creditors challenge the board's decision as a violation of its fiduciary obligations. Whereas, normally, the board's decisions receive the benefit of the business judgement rule so that informed decisions in good faith are unassailable, the courts have created an occasonal exception for decisions made while in the "vicinity of" or "zone of insolvency."[6] The result is a potential shift of fiduciary duties to include creditors as beneficiaries of the board's duties, rather than shareholders alone.

---

[6]*See, e.g.,* Geyer v. Irgensoll Publ'ns Co., 621 A.2d 784, 791 (Del. Ch. 1992) ("[T]he fact of insolvency ... causes fiduciary duties to creditors to arise."); In re NCS Healthcare, Inc., S'holders Litig., 825 A.2d 240, 256 (Del. Ch. 2002) ("[A]s directors of a corporation in the 'zone of insolvency,' the NCS board members also owe fiduciary duties to the Company's creditors."); Credit Lyonnais Bank Nederland, N.V. v. Pathe Commc'ns Corp., Civ. A. No. 12150, 1991 WL 277613, at *34 (Del. Ch. Dec. 30, 1991) ("At least where a corporation is operating in the vicinity of insolvency, a board of directors is not merely the agent of the residue risk bearers, but owes its duty to the corporate enterprise."); *cf* N.A. Catholic Educational Programming Found., Inc., v. Gheewalla, 930 A.2d 92 (Del. 2007) (creditors of corporation in zone of insolvency who are adequately protected by contract clauses do not have standing for direct action about breach of fiduciary duties against its directors).

# 5     Executive Compensation

Options resolve a problem that diversified shareholders have in aligning the incentives of their agents, risk-averse managers. Diversified shareholders tend to approach risk-neutrality with respect to the risks facing the stocks that they own because they are diversified. The good luck of some of their holdings will tend to cancel out the bad luck of others. Thus, from the perspective of diversified shareholders, the managers of those businesses should pursue the projects with the greatest expected returns. For diversified shareholders, sacrificing returns in order to reduce risk makes little sense.

Managers, however, do not have the full benefit of diversification. They only have one career and one employer. Bad outcomes at the firm that they manage likely hurt managers' careers, employment prospects, and wealth. Accordingly, managers tend to display the normal, unmitigated aversion to risk of individuals. The result is a divergence of the interest of agents from those of their principals. The managers, as agents of shareholders, have the incentive to sacrifice returns for safety whereas the shareholders, their principals, would prefer not to sacrifice returns for safety.

Compensating managers with options reduces this discrepancy. Because options increase in value with risk, managers who hold options in their firm should be willing to take more risk than if their compensation were only cash or stock.[7]

---

[7]In the vast literature on executive compensation, evidence that managers take more risk in firms that offer executive compensation with more options appears; *see, e.g.*, Matthew L. O'Connor & Matthew Rafferty, *Incentive Effects of Executive Compensation and the Valuation of Firm Assets*, 16 J. Corp. Fin. 431 (Sep. 2010); *see also, e.g.*, Neil Brisley, *Executive Stock Options: Early Exercise Provisions and Risk-Taking Incentives*, 61 J. Fin. 2487 (2006).

# 6    The Proposal to Restructure by Distributing Options

An interesting application of the usage of options is the proposal to restructure insolvent enterprises by distributing packages of options to the various classes of claimants.[8] The problem is that courts are experts in resolving disputes but not in valuation. Yet, the reorganization process requires both the resolution of disputes between the debtor and creditors (as well as the establishment of priorities between creditors about their claims to specific collateral) and the valuation of the going concern. The option approach relieves courts of the second task. The court needs to merely establish the sizes of the claims of the different classes of creditors. Then, junior creditors receive options that produce value only if the going concern has a value greater than the threshold after which this class would receive value. Successive junior classes receive options written by their immediately senior class. The result should be that the value of the enterprise gets allocated according to priorities without the court having to determine it.

An example illustrates. Suppose the failed enterprise has four classes of claimants. Class A, the secured creditors, has a security interest in all assets of the enterprise and a claim of $100. Class B, the unsecured creditors, has a claim for $75. Class C, the subordinated creditors, has a claim for $50. Class D, the shareholders, claim the residual. The options approach gives the equity of the enterprise to class A but class A must write a call to class B with a strike price of $100 (for the entire enterprise). Class B writes a call to class C with a strike price of $175. Class C writes a call to class D with a strike price of $225. Then on the expiration date, each class decides whether to exercise their call option and get the enterprise, while paying off the senior classes.

---

[8]Lucian A. Bebchuk, *A New Approach to Corporate Reorganizations*, 101 Harv. L. Rev. 775 (1988).

In theory, this is a beautiful outcome. In practice, it requires that the credit markets work very well, since a junior class that seeks to exercise its option to purchase the enterprise likely needs financing to pay the purchase price to its seniors. Yet, if financing to buy the enterprise were easily available, then the reorganization process could have been completed either outside bankruptcy by refinancing or by a workout or, inside bankruptcy, by selling the enterprise as a going concern to a third party. In practice, running the firm, valuing it, and borrowing can be fraught with problems. The same problems that make refinancing or selling the enterprise as a going concern difficult, stymie the options approach to reorganizations.

# 7    Conclusion

This chapter showed two features of the analysis of options. The pricing of options opens the path to valuing more complex financial instruments, explored in Exercises 3–5. The incentives that options produce have an importance that goes beyond pricing, to reach issues such as the shape of executive compensation, fiduciary duties in the zone of insolvency, or the shape of reorganization law.

This is the last, strictly speaking, chapter on quantitative finance. Discounting (Chapter 4), the CAPM (Chapter 5), and the call option valuation formula (this chapter) form the mathematical foundation of finance. The rest of the book explores less mathematical aspects of quantitative policy analysis that touch on financial issues, broadly speaking.

# 8    Exercises

1. Consider that the day is Friday, February 16, 2018, and consider the March 16, 2018, BioMarin Pharmaceutical (symbol BMRN) $85 calls. The stock trades at 84.98 and the risk-free rate is 1.35%. Your analysis indicates that the annual volatility of BioMarin stock is 30%. What is the value of this call according to the Black–Scholes formula?

2. On Friday, February 16, 2018, the March 16, 2018, BioMarin Pharmaceutical (symbol BMRN) $85 calls closed at a price of $3.95. The stock closed at 84.98 and the risk-free rate was 1.35%. What volatility did this price imply for BioMarin? (Hint: If you entered the Black–Scholes formula into Excel for the previous exercise, then a fairly easy way to solve this is to use Excel's Goal Seek feature.)

3. Ford Corporation (symbol F) makes an issuance of a convertible bond. The terms are that each unit will entitle its buyer to a $1000 payment on July 16, 2031, and to annual payments of $35 till then, on July 16 of each year. During early July of 2031, each holder of the convertible bond may choose to receive 50 shares of the company instead of the $1000.

   3.1 Consider the difference of this security from a normal bond. This security implicitly contains a call option. What are the terms of this option and for how many shares is it?

   3.2 What is the value of the option according to the Black–Scholes formula? (Simple version: assume the risk-free rate is 2.4% and that Ford's volatility is 25% and that you are valuing them on February 16, 2018; more complex version: value the option at the time you answer this problem and find the appropriate risk-free rate and volatility.)

   3.3 Using your answers to (or the methods of, if you did not do it) Exercise 3 in Chapter 4, retrieve the continuous interest rate that is implicit in Ford's bonds that are not convertible. Using that rate, value the bond component.

   3.4 What should Ford charge for issuing this security?

   3.5 The date is February 16, 2018, and Ford issued this security in the past. The convertible bond is trading at $1017.46 (and all the other conditions of the simple version of the above exercises hold, i.e., the risk-free rate is 2.4%, the other bonds are priced as in Exercise 3 in Chapter 4, &c). What volatility does this price imply for the stock?

4. A government wants to privatize a state-owned utility company in a way that ensures broad ownership and prevents wealth disparities. Accordingly, they will ensure many small buyers obtain the stock. To prevent wealth disparities, the state will have the right

to buy the stock back at three times the issuing price in five years. Moreover, to ensure that this is appealing to small investors, the buyers of these securities will be entitled to a 3% annual payment for the first five years.

4.1 Describe the components of this security.

4.2 Assume that the stock of utilities like this one has annual volatility of 35%. You expect the stock, if it traded without any of these features, to have a beta of .9 and annual returns of 9%. The stock market has an expected return of 10% and the risk-free rate is 3%. Explain how to price the security that buyers would obtain.

5. BioMarin Pharmaceutical (which we met at the opening exercises) issued on August 7, 2017, convertible subordinated notes at 98% of par which carried an interest rate of 0.599%, were due August 1, 2024, and pay interest on February 1 and August 1 of each year. On their due date, each note of $1000 face value would covert to 8.0212 shares of common stock at the choice of the holder. The company's shares closed at $89.05.

5.1 Assuming volatility of 40% for BioMarin stock, price the option component and calculate the internal rate of return of the bond component.

5.2 Assuming a 5% rate on the bond, calculate the implied volatility of the stock according to the bond's value.

6. The only obligation of a corporation is debt of $50 million that is due next week. If the corporation is solvent, then it can refinance the loan, replacing it with a new one. If the corporation is insolvent, then it will have to file for bankruptcy where the absolute priority rule will apply, meaning that holders of the debt must either receive full value or no claimant junior to them may receive any value; i.e., the holders of the debt will either have their claims satisfied or they will receive the full value of the enterprise. The corporation has outstanding preferred stock with face value of $25 million and a single class of common stock. The absolute priority rule will apply to the preferred as well so that the holders of the common stock may not receive any value unless the holders of the preferred receive full value. State the corporation's debt, preferred, and common stock as

a portfolio of options on the value of the corporation's business or assets (the implicit assumption is that in the bankruptcy, either the business will be sold as a going concern free of any encumbrances and liens or the assets will be sold piecemeal, whichever produces the greatest recovery).

7. Using Mathematica (or any mathematics program) compare the value of a set of call options according to the integration method of Formula 1 to their value according to the Black–Scholes formula, Formula 2.

# 7

# Illustrating Statistical Data

Statisticians recognize that a visual inspection of the data is crucial. Anscombe's quartet is four groups of data that have the same statistical properties and produce the same linear regression yet a visual inspection reveals very different patterns.[1]

Sometimes statistical inference sounds daunting. Yet, besides having some technical tools that utilize probability theory to draw some potentially very powerful conclusions, it is a tool in reasoning and persuasion. A visual representation, however, enhances enormously the persuasiveness of the numbers.

Chapter 5 on the Capital Asset Pricing Model is an example. Consider a description of sensitivity to the market and betas without Fig. 4 in Chapter 5, which shows the actual data points that are the price changes of the stock and the index, and juxtaposes a stock that is highly sensitive to the market and has a high beta, to a relatively insensitive, low-beta stock. The graphical illustration adds not merely explanation but realism.

---

[1] F. J. Anscombe, *Graphs in Statistical Analysis*, 27 Am. Stat. 17 (1973).

© The Author(s) 2018
N. L. Georgakopoulos, *Illustrating Finance Policy with* Mathematica, Quantitative Perspectives on Behavioral Economics and Finance, https://doi.org/10.1007/978-3-319-95372-4_7

Consider, similarly, the presentation of the Capital Asset Pricing Model divorced from Fig. 6 in Chapter 5, which shows not only that returns increase with beta, but also that their standard deviation largely follows the model. Displaying tables of numbers is necessary, but the thrust of a graphical illustration is enormous.

Doubters can always propose alternative causal chains. The author who proposes the thesis bears the persuasive burden of precluding alternative causal chains. In some settings, the statistical tests can perform this function, rejecting alternative hypotheses.

Neither statistical analysis nor the philosophy of logic has an airtight definition of causation; even philosophers concede that causation may be correlation.[2] Arguments about causation necessarily rely on logic. Statistical analysis usually rejects some alternative hypotheses with some confidence; that rejection is merely a tool in the argumentation about causation. An illustration displaying that the data agrees with the reasoning of the theory has great persuasive weight, akin to the way that Fig. 6 in Chapter 5 persuades about the logic of the Capital Asset Pricing Model. Granted, the CAPM is a monumental achievement of theory and evidence, but lesser projects can benefit from graphical illustrations just as much.

# 1    Simple Settings: Plot the Data

In settings where outcomes relate to one or two inputs, the visualization of the actual data is not very complex. The plots of the changes of the price of the stock against the index, Fig. 4 in Chapter 5, p. 62 above, is such an example. This subsection presents some more examples from legal research.

---

[2]The notion that causation is practically indistinguishable from correlation is espoused by some contemporary philosophers; *see, e.g.,* JUDEA PEARL, CAUSALITY: MODELS, REASONING, AND INFERENCE (2d ed. 2009). Causation has posed enormous problems for philosophy, with philosophers taking positions ranging from the notion that causation is a fundamental concept that cannot be reduced to an explanation (the "primitivism" view) to that causation does not exist (the "eliminativist" view); *see* Jonathan Schaffer, *The Metaphysics of Causation,* in THE STANFORD ENCYCLOPEDIA OF PHILOSOPHY (Edward N. Zalta, ed., Fall 2016 Edition), https://plato.stanford.edu/archives/fall2016/entries/causation-metaphysics/ [perma.cc/MTL2-GXBX].

## 1.1     Crime and Abortions

Professors Donohue and Levitt argue that the legalization of abortion led to a dramatic decline in crime, when the unborn cohort would reach the ages of predominant criminality.[3] The premise is that in some states that legalized abortion earlier and in the rest of the United States no later than when *Roe v. Wade* was decided in 1973, women who were not ready to become mothers were able to postpone having children at much larger rates than before. The use of the word *postpone* is important because the authors show that overall fertility did not drop.

Growing up as an unwanted child or a desired child apparently influences the child's criminal proclivities. While the article performs a battery of statistical tests, one of the most telling graphs, Fig. 1 (Figure 4b of the original paper), relates the change of the number of abortions to the change of property crime per person in each state. While the data contain some noise, the relation is undeniable. This display lets the reader see that the overall decline of crime rates do not influence the analysis, all states experience declines in crime. The states where the abortion rates declined more, however, experience greater declines in crime rates.

This Donohue and Levitt research reveals an important lesson about family planning and future criminality. The persuasive power of the paper depends on more than this graph. This graph, however, gives a stark image of the relation between crime and abortions.

The publication appeared in a scholarly economics journal, where measuring changes in crime by their natural logarithm raises no eyebrows. The graph could be more approachable, however, to the reader if the vertical axis bore units that facilitated the direct understanding of the change of crime rather than its logarithm. The lay reader would

---

[3]John J. Donohue III & Steven D. Levitt, *The Impact of Legalized Abortion on Crime*, 116 Q.J. Econ. 379–420 (2001).

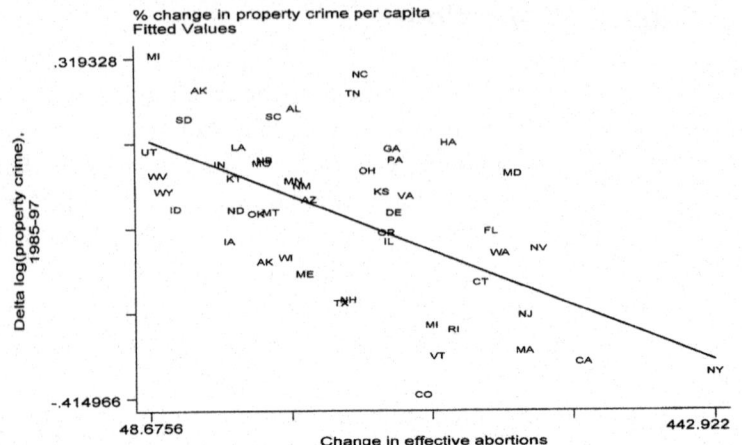

Figure 4b: Changes in Property Crime and Abortion Rates, 1985-1997

**Fig. 1**  Crime and abortion per Donohue and Levitt

rather see the logarithm translated back into crime change, while letting the graph retain the progressive nature of the logarithmic scale.

The key points are that the figure increases the persuasiveness of the argument and that some more friendliness of the figure toward the lay reader would further increase its impact.

## 1.2    Patent Age and Litigation

A second example of the graphical presentation of simple relations appears in a longer article on patent law.[4] The article analyzes various conclusions about the tendencies of patent litigation, one of which is that patent litigation predominantly occurs when a patent is relatively young. The data about patent age and litigation produce Fig. 2.

The horizontal axis holds the age of patents and the vertical axis is the frequency of litigation. Figure 2 completely conveys the findings. Litigation peaks for patents that are three years old and declines

---

[4]John R. Allison, Mark A. Lemley, Kimberly A. Moore, & R. Derek Trunkey, *Valuable Patents*, 92 GEO. L.J. 435 (2004).

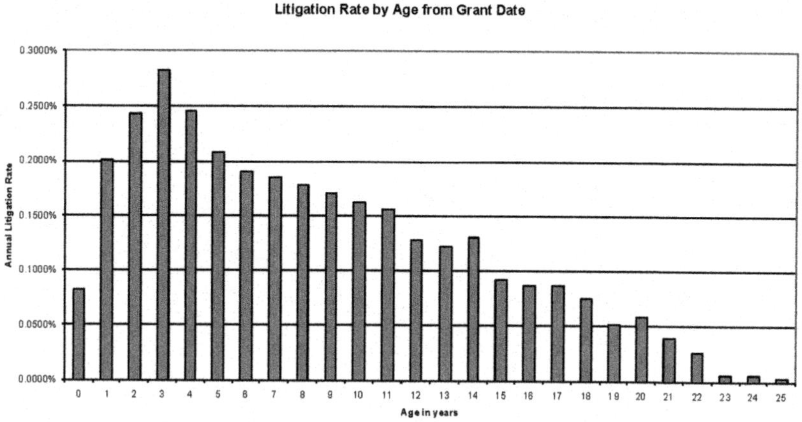

Litigation Rate by Age from Grant Date

Fig. 2    Patent age and litigation per Allison, Lemley, et al.

gradually so that twelve-year-old patents draw half as much litigation. No amount of description, however, can have the clarity of this visual display. The simplicity of the graph is instrumental to its impact.

## 1.3    Grocery Bags and Infections

The last simple example involves an event study, increasing marginally the complexity. The phenomenon on which the study focuses is the imposition by San Francisco of a requirement that grocery shoppers use reusable shopping bags.[5] Obviously, the motivation was environmental. The study, however, focuses on foodborne illness. Figure 3 shows emergency room visits in San Francisco before and after the new requirement. The increase is notable. A different figure in the same paper shows that surrounding counties did not experience a change in hospitalizations, excluding the possibility that a cause that coincided with the adoption of the new measure drove hospitalizations up generally.

---

[5]Jonathan Klick & Joshua D. Wright, *Grocery Bag Bans and Foodborne Illness* (November 2, 2012; U of Penn, Inst for Law & Econ Research Paper No. 13-2), SSRN https://ssrn.com/abstract=2196481.

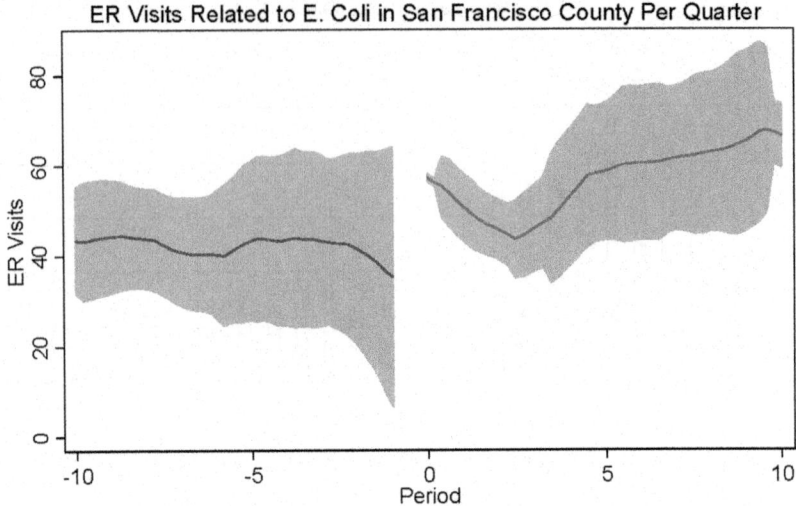

**Fig. 3** Requiring reusable grocery bags and hospitalizations

The figure of emergency room visits around the adoption of the reusable bag mandate introduces a new element. Rather than presenting a single estimate for each month, the figure plots an error band around the point estimates. This is a welcome addition of nuance. The nuance qualifies the interpretation of results that may often appear excessively precise.

Mathematica has a built-in type of plot that includes a display of error bars, **ErrorListPlot**. That, however, produces error bars rather than an error band. A graphic like that of Fig. 3 can be created using the graphic primitive **Polygon**. Using the means and the errors, create a list of coordinates for the points below the means and one for the points above the means.[6] Put the two in a single list while reversing

---

[6]For example, consider that the list baglinebef has its elements in pairs of month and hospitalizations, and that the list bagerrbef has its elements in pairs of month and error values. Create a list adding the error:

```
befpluserr=Table[
        {baglinebef[[j, 1]],
         baglinebef[[j, 2]]+bagerrbef[[j, 2]]}
       , {j, Length@baglinebef}];
```

Also create a list subtracting the errors from the means, using the same method.

one so that the command **Polygon** recognizes the points as being in sequence going around the polygon, while also making the depth of the list consistent.[7] Apply **Polygon** to the result inside a **Graphics** command with appropriate placement of the axes, appropriate specification of tick marks, and a pleasing aspect ratio.[8]

## 1.4    Bar Exam and GPA

Even complex statistical techniques, such as the probit regression, are amenable to graphical illustrations. An unpublished study by this author looked at two years of bar passage data of one law school and related it to that school's GPA and the students' LSAT. Since the outcome variable is binary, pass or fail the bar examination, the linear regression model is inappropriate. Appropriate is a logit or a probit regression method, which tries to explain the data by the proper placement of, respectively, a logistic or a normal distribution. The normal distribution gets located so that the probability it indicates matches the success rate in the data.

The dependent variable was the success of students in the bar examination. The independent variables were the students' grade point average in law school and their score in the standardized test for entrance into law schools, the LSAT. Both independent variables were significant, but the grades of the students were much stronger predictors of their success.

---

[7]Assuming the two lists are befminuserr and befpluserr, join them, reverse the second, and make the depth consistent using Flatten[list, 1]:
```
polygonbef=Flatten[{befminuserr,Reverse@befpluserr},1];
```
The same process produces a set of coordinates to use for creating a polygon for after the change.

[8]For example:
```
Graphics[{GrayLevel[.75],
          Polygon@polygonbef,Polygon@polygonaft,
          Black,Thick,
          Line@baglinebef,Line@baglineaft
    },Axes->True,AspectRatio->.5,AxesOrigin->{-10,0},
  Ticks->{Range[-10,10,5],Range[0,80,20]}]
```

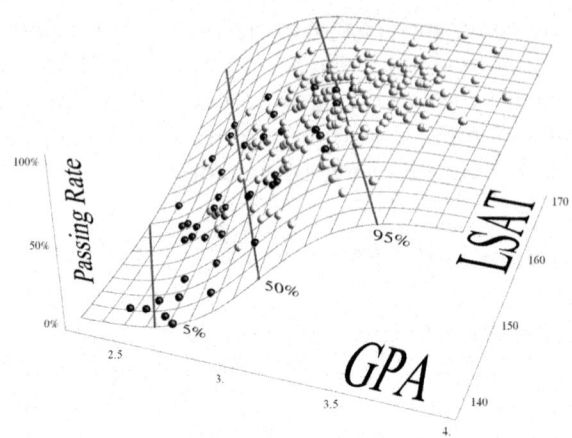

**Fig. 4**  Bar passage prediction and data

The problem with the probit regression is that communicating its results through numbers is not particularly intuitive. The conclusions become much more powerful with a figure that relates LSAT scores, grade point average, and bar passage. Figure 4 makes it immediately apparent that students with low grades and LSAT scores tend to fail the bar and exactly how the probability of passing increases with grades and LSAT.

Figure 4 shows as dark spheres the students who fail the bar and as light spheres those who pass. The spheres rest on a carpet-like surface. The surface is the result of the probit regression, taking values from zero to one corresponding to the normal distribution's cumulative distribution function. The probit regression has related the normal distribution to the LSAT and the GPA of the students so that it comes the closest to explaining their actual rate of passing the bar.

The first step to creating Fig. 4 is the collection of the data. The data takes the form of a list, each element of which has three values, the GPA, the LSAT, and a one or zero corresponding to the bar examination outcome. The command **ProbitModelFit** runs the probit regression and the command **Normal** extracts from the regression the formula for the surface.[9] A **Plot3D** command creates the surface. However, the ranges of the three variables are very different, with LSAT spanning over thirty units from 139 to 170, GPA spanning a little under two units, from 2.3 to 4, and the passing probability spanning one unit from zero to one. While Mathematica scales the three dimensions very intuitively, specifying their ratios using the **BoxRatios** option makes that scaling explicit.[10] That information will become useful when shaping the spheres.

Filtering the data for passing and failing students using the **Select** command produces the corresponding two lists that hold the GPA and LSAT of each.[11] The figure presents each point as a sphere but the **Sphere** command would produce flattened disks because it would use the same radius in every dimension whereas the units scale very differently along each axis. The command **Ellipsoid** takes a separate

---

[9]The code stores the regression in a variable called gpaprob—the data are in the variable bon1. By using g and l as the name of the variables and using the same variable names, we can later apply the formula that the probit regression produces through the command **Normal**:

```
gpaprob = ProbitModelFit[bon1, {g, l}, {g, l}]
```

[10]The code uses g for GPA values, uses l for LSAT values and uses the command **Normal** on the regression, gpaprob. Mathematica offers several ways to manipulate the appearance of the surface; here, the code makes it white by placing white sources of light above it. Also, Mathematica draws a box by default around its three-dimensional figures, and avoiding it requires the option **Boxed** set to **False**.

```
carpet = Plot3D[Normal[gpaprob],
         {g, 2.2, 4}, {l, 139, 170},
         Boxed -> False,
         BoxRatios -> {1, 1, .5},
         Lighting -> {
           {"Point", White, {0, 150, 10}},
           {"Point", White, {0, 150, 2}},
           {"Point", White, {4, 170, 2}}}]
```

[11]The **Select** command relies on a pure function that checks the third element of each datapoint for passage or failure:

```
passing = Select[bon1, #[[3]] == 1&];
failing = Select[bon1, #[[3]] == 0&];
```

radius for each dimension and solves the problem. Setting the ratios of the three dimensions to 1.7, 30, and 2 produces spheres. Those ratios correspond to the range of the values of GPA and LSAT, while the third value is twice the vertical range of the graph, doubled because the height of the box in which the graphic appears is halved. Each student corresponds to a sphere with its $x$-coordinate (width) being GPA, $y$-coordinate (depth) being LSAT, and $z$-coordinate (height) being the value of the normal distribution located according to the probit regression.[12]

The figure also identifies three percentile bands with gray lines. The lines mark the 5, 50, and 95% probability of passing the bar. To create those lines, **FindRoot** obtains the appropriate values of the other

---

[12]The corresponding code first raises all points a little so they are more visible, by zboost. Then it sets the ratios of the dimensions of the ellipsoids through elratios. The results of the probit regression go into two functions, one not raised and one raised by zboost. Then the **Table** command produces a list of the ellipsoids corresponding to each student. The iterator, at the end of each **Table** command, is i and runs from one to the length of each list. Rather than just producing an ellipsoid, a **GrayLevel** appears before each. The color of each point is varied a little, by setting its **GrayLevel** at a random level using the command **Random[Real, range]**, with the range being .8 to .95 for the light gray passing students and 0 to .15 for the dark, failing ones. The ellipsoid radii come from the multiplication of the ratios (in elratios) by a small number that gets the appropriate size.

```
zboost = .01;
elratios = {1.7, 30, 2};
probit[x_, y_] := Normal[gpaprob] /. {g->x, 1->y}
probitzboost[x_, y_] := zboost + probit[x, y]
passrandomgrayspheres = Table[
 {GrayLevel[Random[Real, {.8, .95}]],
  Ellipsoid[{passing[[i, 1]], passing[[i, 2]],
   probitzboost[passing[[i, 1]], passing[[i, 2]]]},
  elratios .012]} , {i, Length@passing}];
failrandomblackspheres = Table[
 {GrayLevel[Random[Real, {0, .15}]],
  Ellipsoid[{failing[[i, 1]], failing[[i, 2]],
   probitzboost[failing[[i, 1]], failing[[i, 2]]]},
  elratios .011]}, {i, Length@failing}];
```

variables at each end of the plot.[13] Those values then go inside **Line** commands to create the percentile lines.[14]

A **Show** command brings together the plot of the surface with the spheres of the data and the lines of the percentiles.[15] The options that are necessary are about having axes, placing them, and about plotting a range that does not omit any parts of the shapes.[16] To liven the result, **Lighting** and **Specularity** specify a more natural lighting and some shininess for the spheres.[17] Finally, the label text is placed in three dimensions through a

---

[13]The code equates each variable to the value of $x$ by applying to $x$ the rule that **FindRoot** produces. The minima appear where the has a value of 139. The maxima appear whenre the has a value of 170 but only for the 50th and 95th percentiles. For the 5th percentile, the appropriate LSAT for the maximum end of the line comes from the minimum GPA:

```
gpa95pcmax=x/.FindRoot[probit[x,170]==.95,{x,2.8}];
gpa50pcmax=x/.FindRoot[probit[x,170]==.5,{x,2.8}];
lsat05pcmax=x/.FindRoot[probit[2.3,x]==.05,{x,140}];
gpa95pcmin=x/.FindRoot[probit[x,139]==.95,{x,2.8}];
gpa50pcmin=x/.FindRoot[probit[x,139]==.5,{x,2.8}];
gpa05pcmin=x/.FindRoot[probit[x,139]==.05,{x,2.8}];
```

[14]Each line has the corresponding GPA, LSAT, and probability as the coordinates of each end of the line:

```
percentilelines={
 Line[{{gpa50pcmax,170,probit[gpa50pcmax,170]},
  {gpa50pcmin,139,probit[gpa50pcmin,139]}}],
 Line[{{gpa95pcmax,170,probit[gpa95pcmax,170]},
   {gpa95pcmin,139,probit[gpa95pcmin,139]}}],
 Line[{{2.3, lsat05pcmax,probit[2.3,lsat05pcmax]},
  {gpa05pcmin,139,probit[gpa05pcmin,  139]}}]}
```

[15]The basic code would be:

```
Show[carpet,Graphics3D[{
        passrandomgrayspheres,
        failrandomblackspheres,
        percentilelines}]]
```

[16]Which means inserting at the last position of the **Show** command:

```
,PlotRange ->All,
Axes ->True,
AxesEdge ->{{-1, -1}, {1, -1}, {-1, -1}}
```

[17]Thus, before the passing spheres, the code is:

```
Lighting ->"Neutral", Specularity[White, 50],
```
Whereas before the dark spheres the code makes them less shiny:
```
Specularity[White, 20],
```

function that converts data for textual display into a 3D object that can be placed at will.[18]

## 2     Complex Settings: Plot the Residuals

The preceding figures have the simplicity of juxtaposing only a number of phenomena that can be visualized, i.e., two in the cases of crime against abortion, litigation rates against patent age, or hospitalizations against grocery bag laws, and three in the case of bar passage rate against GPA and LSAT. Often, the statistical analysis has greater complexity, relating an outcome to several inputs, usually with a linear multiple regression. When several of the explanatory factors have significant effect on the outcome variable, plotting the data itself severely understates the strength of the relation between the output variable and the several input variables. If the graph, for example, displays the crime rate in each city against the city budget, two similar budgets may correspond to different crime rates, but a different variable may explain that difference. The raw data may give the appearance of a discrepancy, but the model may explain it.

---

[18]The complex function is part of the notebook's initialization:

```
text3D[str_, location_: {0, 0, 0},
   scaling : {_?NumericQ, _?NumericQ} : { 1, 1},
   longitude : _?NumericQ : 0, elevation : _?NumericQ : 0,
   tilt : _?NumericQ : 0, opt : OptionsPattern[]] :=
Module[{mesh =
DiscretizeGraphics[
  Text[Style[str, opt, FontFamily -> "Times",
    FontSize -> 12(*don't scale with ftsz*)]], _Text,
  MaxCellMeasure -> 0.04],
  rotmatrx =
  RotationMatrix[
    longitude, {0, 0, 1}].RotationMatrix[-elevation, {0, 1,
      0}].RotationMatrix[tilt, {1, 0, 0}]},
  Join[{EdgeForm[]},
  MeshPrimitives[mesh,
2] /. {x_?NumberQ,
y_?NumberQ} :> (rotmatrx.(scaling~Join~{1} {x, y, 0}) +
location)]]
```
Then the **text3D** function can be deployed in the command creating the graph.

Despite that raw data obscures the relation of the dependent variable to each independent variable, figures illustrating the relation between the variables are possible. Rather than displaying the raw data, the figure would display the residual, the variation of the actual data from the expected value, i.e., the expected relation according to the statistical model. Returning to the two cities of the example, suppose the third variable that explains the difference is one that captures whether the city has significant gambling, as would Las Vegas and Atlantic City. Taking into account this extra variable makes the two cities that have different crime rates have expected crime rates equal to the crime rates that they actually have. The model explains both crime rates accurately. Because the model explains both crime rates, therefore both cities will appear on the line of the model's prediction in a graph of the residuals.

An example of such figures appears in a study of the resources that different countries devote to securities enforcement.[19] Professors Jackson and Roe examine several characteristics of the legal system of each country: disclosure obligations, standards for liability, shareholder rights against the entrenchment of directors. They also control for the countries' wealth and judicial efficiency. All those factors and the staffing level of the securities regulator are the independent (input) variables used to explain various measurements of capital market vitality. Figure 5 plots the residual, in other words the error from the model's prediction, of the ratio of each country's number of listed firms to its population on the vertical axis against the staffing level of its securities regulator on the horizontal axis. The model accounts for all the other variables, and the figure lets the reader observe only the relation between staffing and listed firms. Again, some noise exists, but the figure makes the relation unmistakable.

The graph makes the point that staffing determines stock market vitality much more clearly than tables of figures. After adjusting for all the other features of each legal system, the graph shows that the staffing

---

[19] *See generally* Howell E. Jackson & Mark J. Roe, *Public and Private Enforcement of Securities Laws: Resource-Based Evidence*, 93 J. FIN. ECON. 207 (2009).

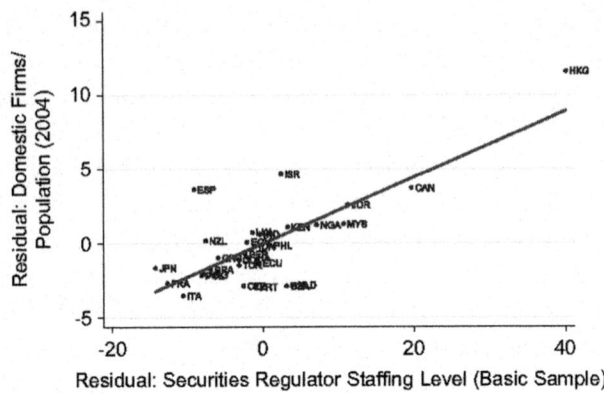

**Fig. 5**  Residuals of listed firms against regulator staffing

level of the securities regulator has a strong relation to the number of each country's firms that are listed in the stock market.

Figure 5 is a type of graph called "added variable plot" or "partial regression plot." The dependent variable, on the vertical axis, has a predicted level depending on the values of the other variables. Thus, one would expect that the horizontal axis would hold the actual value of one of the other variables, which is certainly an acceptable alternative presentation of the data. However, the variables that are not being displayed, such as wealth or judicial efficiency in this research, may have a relation with the variable that lies on the horizontal axis. Perhaps wealthier countries, all else equal, have a disproportionately greater staffing of their securities regulators. This type of graph takes those relations into account, so that the values of the other variables relate to the values of the horizontal axis. In a sense, as the values change along the horizontal axis all else is not equal but changes according to the relations observed in the data.[20]

---

[20] *See, e.g.,* Tom Ryan, Modern Regression Methods (1997); John Neter, William Wasserman, & Michael H. Kutner, Applied Linear Statistical Models (3rd ed. 1990); Normal R. Draper & Harry Smith, Applied Regression Analysis (3rd ed. 1998); R. Dennis Cook & Sanford Weisberg, Residuals and Influence in Regression (1982); David A. BelsleyEdwin Kuh, & Roy E. Welsch, Regression Diagnostics (1980); and Paul F. Velleman & Roy E. Welsch, *Efficient Computing of Regression Diagnostics*, 35 Am. Stat. 234–42 (1981).

The plot is the result of two regressions. First, regress the variable displayed on the horizontal axis against the other independent variables, and obtain the residuals for each observation. Those residuals lie on the horizontal axis. Then, regress the dependent variable of the vertical axis on all the other variables, not including the one on the horizontal axis, and obtain the residuals for each observation. Those go on the vertical axis, forming the added variable plot.

Counterintuitively, both axes have zero values where the residuals are zero, rather than where the underlying variables are zero. Return to Fig. 5. The location of Hong Kong at the far right and of Japan at the far left means that Hong Kong's securities regulator is staffed much more than the other variables explain and that Japan's is staffed much less. If Japan and Hong Kong were both at zero, this would not mean that they staff their regulators equally but that they staff their regulators exactly as their other attributes predict that they would.

The fact that several regressions are necessary for the production of a plot like Fig. 5 means, in Mathematica, a few extra lines of coding. The data must be rearranged to the appropriate input order, which means creating two new lists of the appropriate parts of the data with the variable to be regressed against the others at the last position.[21] The regressions are applications of the **LinearModelFit** command.[22] Then, a

---

[21]Accordingly, for the horizontal axis, the variable to go on the horizontal axis must go last and the dependent variable must be absent. For example, assume the data is stored in the list datapts and corresponds to a judicial efficiency score, the staffing value, a wealth value and a value for the firms per population. The new list needs to take each element of datapts and have its first element, its third element, and its second element, in that sequence. The corresponding command uses the abbreviation for **Take**, the double square brackets, and the double semicolon to span over all elements of datapts; then, in the second position inside the square brackets is the list of elements of each datapoint to take to form the new list of data:

```
horizdata=datapts[[;;,{1,3,2}]]
```

The data for running the regression for the vertical axis, then, needs to have the dependent variable in the last position and must omit the second item of each datapoint, the number of firms:

```
vertdata=datapts[[;;,{1,3,4}]]
```

[22]To wit:

```
hrgn=LinearModelFit[horizdata,{x1, x2},{x1,x2}]
vrgn=LinearModelFit[vertdata, {x1, x2},{x1,x2}]
```

new list of the corresponding residuals forms the points for the plot.[23] The line of the plot requires a third regression, a regression of those residuals.[24] The **Graphics** command displays the results using the graphics primitives **Point** and **Line**.[25] The location of the axes flows from the option **AxesOrigin**.[26]

# 3    Conclusion

Statistical analysis derives patterns of relationships in data. Yet, our own eyes are extraordinary at recognizing patterns. The visual presentation of statistics has enormous value. This chapter reviewed several ways to visualize data and statistical analysis.

---

[23] The residuals are the actual values minus the predicted values:
```
pltpts=Table[{
  horizdata[[j,3]]-Normal[hrgn]/.
  {x1->horizdata[[j,1]],x2-> horizdata[[j,2]]},
  vertdata[[j,3]]-Normal[vrgn]/.
  {x1->vertdata[[j,1]],x2->vertdata[[j,2]]}
  },{j,Length@horizdata}]
```

[24] Namely,
```
nline=LinearModelFit[pltpts,x,x]
```
As a matter of verification, the slope produced by this regression must match the coordinate of the corresponding variable in the regression of the dependent variable against all the independent variables.

[25] The corresponding command forms the line using the **Table** command and having the iterator j take only two specific values: the minimum and the maximum of the x-coordinates in pltpts:
```
Graphics[{
        Point@pltpts,
        Point@pltpts,
        Line@Table[
        {j,Normal[nline]/.x->j},
        {j,{Min@pltpts[[;;,1]],Max@pltpts[[;;,1]]}}]
  },Axes->True]
```

[26] The corresponding option could be:
```
AxesOrigin->{-1,-1.5}
```

# 4     Exercises

1. The opening paragraph mentioned Anscombe's quartet, the four sets of data that have the same statistical properties and produce the same linear regression but a visual inspection reveals them to have very different patterns. Below appear the four sets of data. Give their descriptive statistics and run a linear regression on each set. Then produce the graph for each. What is your explanation about the mechanism that created each set? (Table 1).

2. Two types of charts that tend not to be used in scholarship but are used a lot in business are the pie chart and the bubble chart. Read the corresponding help pages of the Mathematica documentation and discuss their advantages and drawbacks.

**Table 1**     Data for Anscombe's Quartet

| I | | II | | III | | IV | |
|---|---|---|---|---|---|---|---|
| x | y | x | y | x | y | x | y |
| 10.0 | 8.04 | 10.0 | 9.14 | 10.0 | 7.46 | 8.0 | 6.58 |
| 8.0 | 6.95 | 8.0 | 8.14 | 8.0 | 6.77 | 8.0 | 5.76 |
| 13.0 | 7.58 | 13.0 | 8.74 | 13.0 | 12.74 | 8.0 | 7.71 |
| 9.0 | 8.81 | 9.0 | 8.77 | 9.0 | 7.11 | 8.0 | 8.84 |
| 11.0 | 8.33 | 11.0 | 9.26 | 11.0 | 7.81 | 8.0 | 8.47 |
| 14.0 | 9.96 | 14.0 | 8.10 | 14.0 | 8.84 | 8.0 | 7.04 |
| 6.0 | 7.24 | 6.0 | 6.13 | 6.0 | 6.08 | 8.0 | 5.25 |
| 4.0 | 4.26 | 4.0 | 3.10 | 4.0 | 5.39 | 19.0 | 12.50 |
| 12.0 | 10.84 | 12.0 | 9.13 | 12.0 | 8.15 | 8.0 | 5.56 |
| 7.0 | 4.82 | 7.0 | 7.26 | 7.0 | 6.42 | 8.0 | 7.91 |
| 5.0 | 5.68 | 5.0 | 4.74 | 5.0 | 5.73 | 8.0 | 6.89 |

# 8

# Probability Theory: Imperfect Observations

## 1 Introduction

The previous chapters already engaged probability theory. Here, probability theory reappears ostensibly as a question about evaluating an uncertain signal, imperfect testimony, or an imperfect DNA test. Most signals, testimonies, and tests have imperfections. Understanding how to interpret them is crucial. The probability theory involved, however, can get very complicated. Visualizations often make the setting understandable.

A more fundamental reason for this chapter is closer to finance. Voluminous statistical evidence indicates that the financial markets are "efficient," meaning that they process new information accurately. Going back to the evidence of the CAPM, the stock market appears to have been accurate with respect to knowledge that was still undiscovered. Whereas the CAPM was found and published in the sixties, the market operated according to the CAPM well before! The statistical paradoxes presented in this chapter seek to have a sobering effect. If such counterintuitive settings arise, will the market process the

© The Author(s) 2018
N. L. Georgakopoulos, *Illustrating Finance Policy with* Mathematica,
Quantitative Perspectives on Behavioral Economics and Finance,
https://doi.org/10.1007/978-3-319-95372-4_8

corresponding information correctly? The evidence suggests mostly yes, despite the fallibility of individuals, us, who may well be misled by these paradoxes.

## 2    When to Add and When to Multiply

Uncertainty usually has several sources. A puzzle for the uninitiated is how to derive the final probability from the probabilities of the individual uncertainties, when those can be estimated. Confusion can arise because sometimes probabilities must be added, although usually probabilities must be multiplied.

Depart momentarily from using coin tosses to understand probability puzzles, and use dice. A normal, six-sided die has a probability of 1/6th, or 0.167, of producing each of its outcomes, one through six. Suppose that the player wins in two of those outcomes, say three and five. How do we calculate the player's chances of winning? Here, the appropriate action is to add the probability of one outcome, the three, say, which is 1/6th, to the probability of the other outcome, the five, say, which is also 1/6th, and get 2/6ths or 1/3rd. It is appropriate to add the probabilities of the individual outcomes when they result from a single source of uncertainty (as in this single roll of one die) and the combined outcome arises from either individual outcome.

Consider that the player wins if one number occurs and then, at a second cast of a die, another number occurs, say a three then a five. Now the appropriate action is multiplication. The first number occurs a sixth of the time and only if that occurs then a sixth of that time the second number will arise and the player will win. The player wins with probability 1/36th, the product of the two probabilities. When different uncertainties must produce specific outcomes for a combined outcome to obtain, the probabilities of the specific outcomes must be multiplied.

A misleading setting arises when the game states that the player wins if one coin toss is heads or a second coin toss is tails. One might be misled by the "or" to add the two probabilities of 50%, but one must not add .5 plus .5 to get 100%, although in this setting the error becomes obvious because the player cannot win 100% of the time. Addition is proper because a win results from several outcomes. Addition, however, is the false action because the sources of uncertainty are two, the two tosses. Specify the winning circumstances with greater detail. The player wins if the first toss is heads. The player also wins if (a) the first toss is tails and (b) the second toss is tails. These two paths to winning should now be added to obtain the complete chance of a win. The first is .5 and the second is the product of .5 by .5, or .25. Therefore, the chance of a win is .75.

In actual settings, often these calculations are confounded by risks being related or, in the language of probability theory, not *independent*. Two risks are independent if the outcome of the second does not tend to also occur or not occur depending on the outcome of the first. Examples of risks that are not independent are numerous. Suppose the geology of underground faults is such that pressure builds over centuries, but an earthquake at one position of a fault that has a length of many hundreds of miles releases that pressure; then, an earthquake in one region of the fault reduces the risk of an earthquake at a different region. Vice versa, if the geology of faults is such that one movement precipitates more (as in aftershocks), then an earthquake at one region of a fault would increase the probability of earthquakes in different regions of the same fault. Weather phenomena are also related. Two nearby cities each have some probability of a weather event, but their experiences are related because they are likely to flow from the same event, the same weather system. In the discussion of the Capital Asset Pricing Model (Chapter 5), we saw that a key understanding of that theory is that, while all businesses have some risks that are independent, all businesses also depend on the performance of the overall economy, doing relatively better in booms than in recessions. Regional economic performance has similar characteristics, with local booms tending to help businesses in the region and, vice versa, local recessions tending to hurt.

# 3    The Rare Disease Paradox

One of the paradoxes of probability theory is the rare disease test. The disease is very rare, infecting, for example, one percent of the population. The test is fairly accurate, for example at the 90% level. For

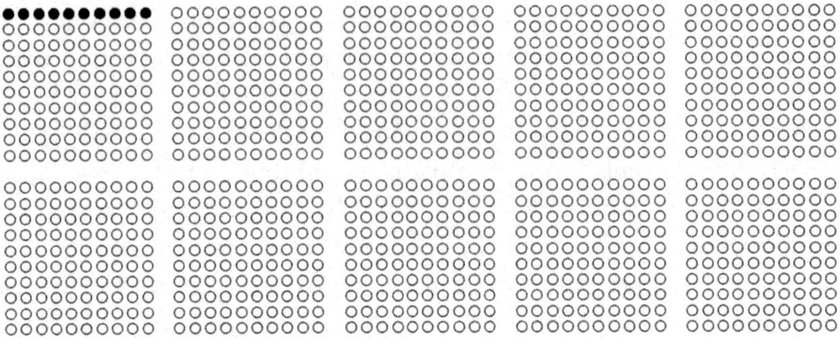

**Fig. 1**   Rare disease: population

simplicity, that single descriptor of the test's accuracy applies both when the test finds the disease and when it reports its absence. Thus, the test will fail to identify as diseased ten percent of the diseased population and will fail to identify as disease-free ten percent of the healthy population.

The paradox asks the probability that an individual who receives a positive test truly has the disease. The lay intuition gravitates toward the accuracy of the test and estimates the subject's probability of being diseased near the ninety percent of the test's accuracy. The truth is very far, however, with the true probability below nine percent.

To see the true probabilistic calculation, start by considering the real image of a population of a thousand. The dark circles are those who have the disease and the light circles do not. Figure 1 shows the reality of the disease in the population. Ten dark circles in a field of a thousand points. The important point at this stage is that nobody actually can observe this reality. Applying the test to this population will produce a different picture.

To observe the application of the test to the population, keep track of both types of errors. The test will err ten percent of the time when applied to the diseased and ten percent of the time when applied to the healthy. Figure 2 demonstrates the application of the test by changing the color of one random point in each row of ten points. One of the truly diseased points becomes white. Ninety-nine of the truly healthy points become dark.

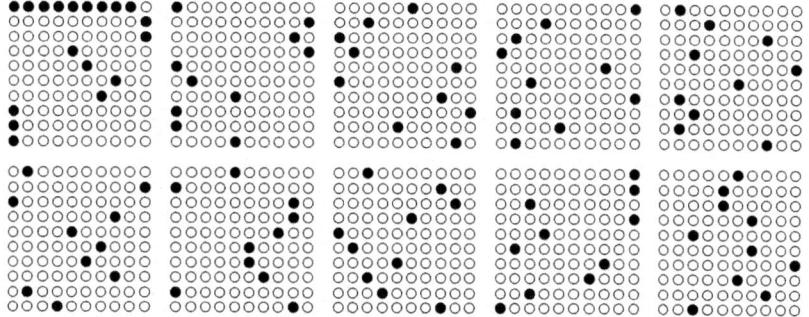

**Fig. 2** Rare disease: test application

Observing Fig. 2 should reveal the content of the information that a positive test carries. The information that the test is positive means that we have landed on one of the dark points of the figure. Since each uninfected row has a false dark spot, that produces 99 falsely dark points. Of the ten infected people, nine correctly dark ones remain. In other words, the figure has 108 dark points and only nine of those truly have the disease. The probability that the subject has the disease is nine in 108, well under 9%.

Moving back from imaging to probability calculations starts by producing the four groups, the probability weight of (i) infected who receive a positive test, (ii) infected who receive a negative test, (iii) healthy who receive a negative test, and (iv) healthy who receive a positive test. Each is the result of a multiplication of the probability of the disease and the accuracy of the test, or one minus their values. (a) Ninety percent of the diseased receive positive tests, and they are one percent of the population. The multiplication is the accuracy rate times the rate diseased. (b) Ten percent of the diseased receive negative results, and the diseased are one percent of the population. The multiplication is one minus the accuracy times the rate diseased. (c) Ninety percent of the healthy receive negative results, and they are 99% of the population. The multiplication is the accuracy rate times one minus the rate diseased. (d) Finally, ten percent of the healthy receive positive tests, and the healthy are 99% of the population. The multiplication is one minus the accuracy rate times one minus the rate diseased.

A verification of the calculation so far should show that all these probability weights sum to one, to a hundred percent. This ensures that the analysis accounted for every group in the population.

Once these four groups are measured, we can answer questions about the accuracy of a positive and a negative test. The questions are (1) the probability that a positive test is correct; (2) the probability that a negative test is correct; as well as (3) the probability that a positive test is false; and (4) the probability that a negative test is false.

In each case, the answer is a matter of selecting the appropriate groups for the denominator and the appropriate one of them for the numerator of the fraction that gives the probability. The two groups that form each test outcome form the sum that belongs in the denominator. In the case of a positive test those would be the diseased who correctly receive positive tests and the healthy who erroneously receive positive tests (multiplications a plus d). If the question is the probability that a positive test is accurate, the numerator must be the accurately positive ones (multiplication a).

# 4    The Cab Color Paradox

The rare disease test is sufficiently complex that an additional example is worth observing. The cab color paradox has a witness testifying about the color of a cab involved in an accident. The circumstances are such that the accuracy of the witness is imperfect and the witness identifies the cab color as not belonging to the majority of the cabs.

For example, suppose the cabs of this city are 80% dark, else light colored and that under the circumstances the witness is accurate 75% of the time. When the witness testifies that the color of the cab was light, what is the probability that indeed the cab was of light color? The lay intuition gravitates toward the 75% accuracy of the witness. The true probability of accuracy is under 50%.

Again, supposing the city has a thousand cabs, Fig. 3 illustrates them as dark and light points. This is the reality, which will be filtered through the witness's observation.

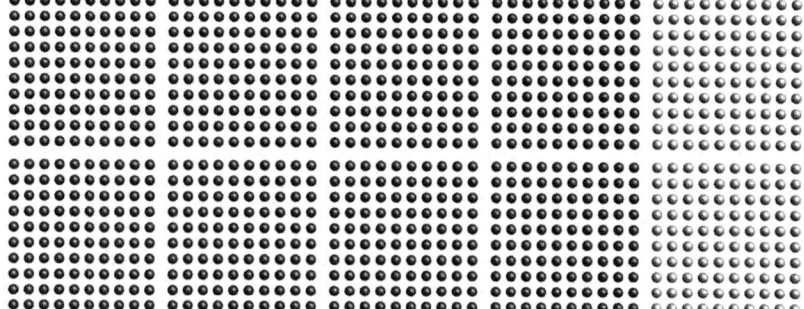

**Fig. 3**  The actual cab colors

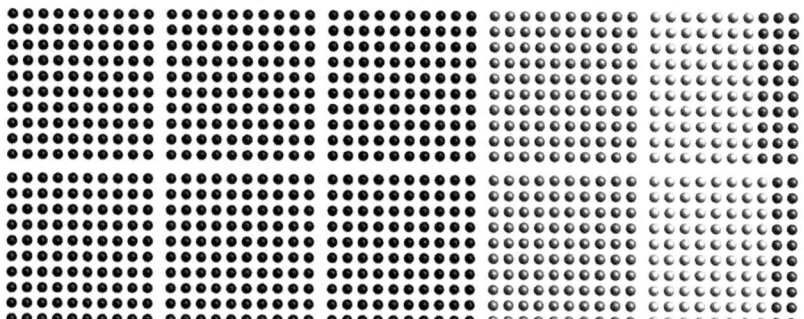

**Fig. 4**  Cab identifications by the witness

The witness errs in identifying both groups, producing a set of four categories: (a) dark cabs called dark; (b) dark cabs called light; (c) light cabs called light; and (d) light cabs called dark. Figure 4 demonstrates these four groups as points having four different levels of darkness, although the witness only separates two groups.

The four groups are the results of analogous multiplications as in the rare disease example. (a) With probability 75%, the witness correctly identifies as dark the dark cabs, which are 80%. This is the first group of points in the figure, the darkest points. (b) With probability 25%, the witness falsely identifies as light the dark cabs, which are 80%. This is the second group of the figure, the medium gray points. (c) With probability 75%, the witness correctly identifies as light the light cabs, which are 20%. This is the third group of the figure, the white points. (d) Finally, with 25% probability, the witness identifies as dark the light cabs, which are 20%. This is the last group of points in the figure, the small group of dark gray points on the right.

Since the witness testifies that the cab is light, the relevant two groups are the middle ones, the two groups called light by the witness. Those will form the denominator of the fraction that will give the accuracy of the statement. The two relevant groups also appear in Fig. 5, where the first and fourth groups are missing, leaving only the two groups of cabs seen to be light. The numerator is the accurately light ones, the set of white points. That is visibly smaller than the first group. The testimony means having landed on one of these points. We have landed on a truly light point with less than 50% probability.

**Fig. 5** The relevant groups when the testimony is "the cab was light"

The application of the probability weights to a number of cabs, as in the example of the city with the thousand cabs, may help the visualization but it is neither necessary nor done in practice. The true calculation uses probability weights, which disentangles the calculation of the probability from issues about visualizing fractional cabs.

The calculations of both the rare disease and the cab color paradoxes can also form a table (Table 1). The top panel of the table holds the probabilities of the different states. Call the state of having the disease or being a light colored cab positive and that of not having the disease or not being a light colored cab negative. The four states are (a) negative correctly identified as negative; (b) negative falsely identified as positive; (c) positive correctly identified as positive; and (d) positive falsely identified as negative. Those are defined by different products of the probability of being positive, $p_p$, and the probability of correct identification, $p_c$, or one minus those probabilities. Those four probabilities must add to one as a matter of verification of the calculations.

The second panel of the table calculates the probability of each statement, signal, or identification. The four statements are (1) a true positive; (2) a false positive; (3) a true negative; and (4) a false negative. The

**Table 1**  States and statements: probabilities

| States | |
|---|---|
| (a) Negative called negative | $(1-p_p)p_c$ |
| (b) Negative called positive | $(1-p_p)(1-p_c)$ |
| (c) Positive called positive | $p_p p_c$ |
| (d) Positive called negative | $p_p(1-p_c)$ |
| | Verification: sum $=1$ |
| *Accuracy of statements* | |
| (1) True positive | $\dfrac{p_p p_c}{p_p p_c + (1-p_p)(1-p_c)}$ |
| (2) False positive | $\dfrac{(1-p_p)(1-p_p)}{p_p p_c + (1-p_p)(1-p_p)}$ |
| | Verification: sum $=1$ |
| (3) True negative | $\dfrac{(1-p_p)p_c}{(1-p_p)p_c + p_p(1-p_c)}$ |
| (4) False negative | $\dfrac{p_p(1-p_c)}{(1-p_p)p_c + p_p(1-p_c)}$ |
| | Verification: sum $=1$ |

verification here works in pairs. The true positive and the false positive must add to one. The true negative and the false negative must add to one. Creating a spreadsheet with the proper formulas is a helpful tool.

A more visual aid to this probabilistic calculus is the probability tree. The resolution of each uncertainty is a node from which as many branches extend as are the possible outcomes. In these examples, the uncertainties are two. The first uncertainty is the determination of the state. Is the individual diseased? Is the cab light colored? Each has two outcomes, positive and negative, hence two branches extend from the first node. With probability $p_p$ the state is positive and the rest of the time, $1-p_p$, negative. The second uncertainty is the determination of the signal. Is the test accurate? Is the witness correct? Again, each has two outcomes, true and false. Therefore, two branches extend from each signal node. With probability $p_c$ the signal is correct or true, and the rest of the time, $1-p_c$, the signal is false. This produces the four outcomes. The probability of each outcome is the product of multiplying the probabilities along the branch leading to it. This produces the four probabilities of the states (a) through (d). To determine the accuracy of the two signals and their inaccuracy, one must continue to form the fractions (1) through (4).

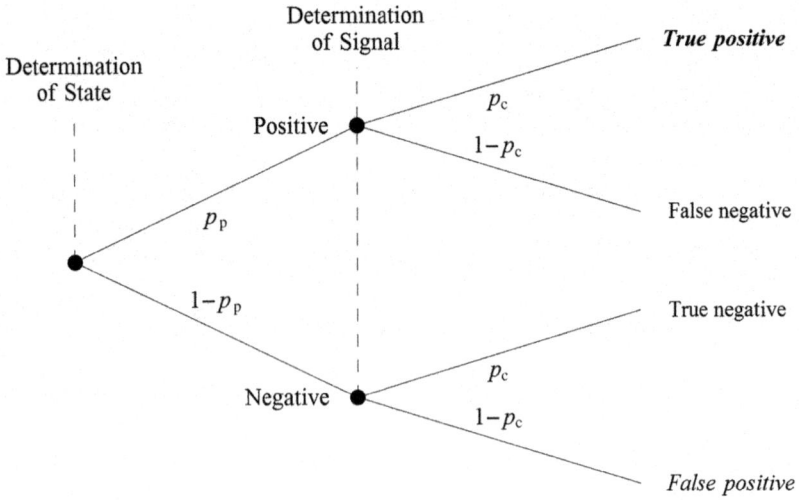

**Fig. 6**  Simple probability tree

Because the examples ask the probability that a positive statement is true, the figure has the relevant outcomes, the positives, in italics. Their sum is the denominator of the fraction that gives the probability that a positive statement is true. The true positive is also in bold to identify it as the numerator of that fraction (Fig. 6).

In sum, the visual approaches discussed here help track the numerical calculations necessary to derive correct probabilistic conclusions. The next sections apply these lessons to two topics, the evaluation of stale DNA and the consequences of judicial and prosecutorial imperfection.

# 5    The Evaluation of Weak DNA Evidence

The evaluation of DNA evidence is usually straightforward in the sense that a single suspect exists and DNA evidence merely supports other evidence with the advantage of extraordinary precision. However, in some instances, the only evidence implicating the suspect is DNA. The quality of the DNA is degraded, perhaps due to the sample being old or small and not allowing the very high precision that a good sample provides. Then, the probabilistic calculus ends up suggesting guilt with some probability. Establishing that probability requires an assessment that has similarities with the preceding paradoxes about the rare disease or the cab color. However, the corresponding analysis is more complex.

The example analyzed here uses the facts from a 1972 rape-murder in San Francisco, California, well before DNA technology. However, the police maintained samples of the likely perpetrator's blood and eventually had its DNA sequence extracted. California also started maintaining a database with the DNA of all sex offenders. In 2003, by scanning the database, California prosecutors identified as a suspect for the murder John Puckett.[1] The one suspect from 1972, Baker, who was investigated but not prosecuted, died without leaving a DNA sample.

The only evidence against Puckett was the DNA identification, which also identified the perpetrator as Caucasian. The prosecutor claimed

---

[1]People v. Puckett, No. SCN 201396 (Cal. Super. Ct. Feb. 4, 2008). The analysis here parallels that in N. Georgakopoulos, *Visualizing DNA Proof,* 3 Crim. L. Prac. 24 (2015–2016).

that the probability of Puckett's guilt was the accuracy of the DNA identification, implying an error rate of less than a ten-thousandth of one percent. The DNA identification had an error rate of about one in 1.1 million, much less accurate than most DNA identifications. The age of the sample meant that a smaller than usual sequence had been read. The defense for Puckett based its argument on the application of the test to a population with the size of the database, about 338 thousand individuals. The probability that the test correctly rejected all of them, assuming that they were all innocent, was under 75%. The defense argued that Puckett could be the one unlucky member of a set of innocents that might be identified with over 25% probability as guilty.

The defense's reasoning relies on the multiplicative nature of repeating an uncertainty. A similar effect appeared in the analysis of diversification, at pp. 56–60 above, where the probability of winning or losing all the tosses dropped dramatically as the number of tosses increased because of the necessary chain of tosses: half the time of the first toss, and half the time of second toss, making it a quarter of the time, and so on. To make tracking the numbers easier, assume the accuracy of the DNA identification were 99%. Then, correctly rejecting the guilt of the first person in the database would happen 99% of the time, and 99% of that would also see the second correct rejection of guilt, and so on. The defense's argument is that correctly rejecting the guilt of 338 thousand individuals means multiplying the accuracy of the test 338 thousand times. Even using the prosecution's estimate of its accuracy, the result is 99.99990909% raised to the power of 338 thousand, which is 73.54%. The probability of correctly rejecting all members of the database was under 74%.

Neither the prosecution nor the defense is correct. The correct answer is not easy and follows a complex probabilistic analysis that can serve as an example for such cases. Just as in the rare disease and the cab color problems, we need to identify a numerator that corresponds to correct instances of DNA identification and a denominator that corresponds to the sum of all possibilities leading to positive identifications, including false ones. In the visual example, we are told we landed on a dark point and we need to calculate the probability of it truly being dark. We need to calculate the probability of obtaining a DNA identification despite that Puckett is innocent as well as the probability of obtaining it if he is guilty.

The path of probabilistic events that leads to a true positive identification is obvious: Puckett is the perpetrator, is in the database, and a DNA match occurs. False positives, however, can arise from several paths. First, the alternative suspect, Baker, might be the true perpetrator and Puckett gets a false positive. Second, some unknown person, not in the database, is the perpetrator and Puckett gets a false positive. Finally, the true perpetrator may be in the database, gets a false negative, and then Puckett gets a false positive.

A probability tree may clarify the road ahead. The first branching may be the least intuitive. It seems that one of three states may occur. The perpetrator is either (a) the alternative suspect, Baker; (b) an unknown not in the database; or (c) someone in the database. More appropriate, however, is to place first the duality that the perpetrator either is the alternative suspect, Baker, or not. Then follows the second branching of the perpetrator being not in the database or in the database. In the first two instances, the next branching is that either a plausible suspect like Puckett receives the only false positive or not, which means that the false positives are either zero or two or more.[2] In the last branching, if the perpetrator is in the database, then the perpetrator either receives a true positive or a false negative. After a true positive, again, either no other plausible suspect receives a (false) positive or more false positives appear. If the perpetrator receives a false negative, then either one plausible suspect receives a false positive or not, in which case

---

[2]Whether Puckett's identification by the database corresponds to exactly one positive identification or to one or more is an important and potentially contested matter. If the entire database was searched and prosecution would only follow a single positive identification, then proceeding according to the analysis of the text, that the identification corresponds to a single positive, is correct. If the search for a match stopped upon reaching the first positive identification, or if the appearance of two or more positives would then lead to further investigation of alibis to bring them down to one who would be prosecuted, then the proper approach is to consider that the identification corresponds to one or more positives. With the actual parameters of this setting, using the analysis that Puckett's identification corresponds to one or more positives would produce a trivially different result. In different factual settings, however, especially if the database had a much greater size, then the latter method could produce a greater probability of a false positive and lower probability of true guilt for Puckett.

the false positives are zero or two or more. Figure 7 shows this probability tree.

The probability tree of Fig. 7 seeks to demonstrate the paths to the relevant outcomes. The branches do not have their probabilities because we do not know them; the analysis that follows will derive them. To restate the structure of the tree, the first uncertainty is about the probability that the alternative suspect that was investigated at that time, Baker, was actually the perpetrator. If one is confident that the police would not cease pursuing a likely suspect, that would be a low probability. The next uncertainty is whether the perpetrator would be in the database. If one is confident that criminals are recidivists and are eventually apprehended, then that probability is high. The next uncertainty is about the perpetrator who is in the database receiving a true positive. That is the accuracy of the DNA identification. These alternatives get to the final nodes of each branch. There, the tree needs to account for the fact that only one positive is observed. That depends on the accuracy of the DNA identification to exclude innocents and the size of the relevant population of the database.

To calculate the corresponding probabilities, an accurate database size is necessary. The estimation of that comes from comparing 1972 San Francisco to the 2003 database of those convicted of sexual crimes.

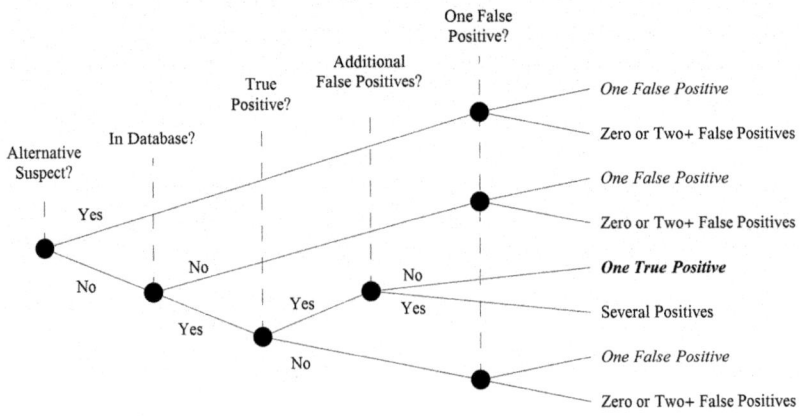

**Fig. 7** Puckett's probability tree

## 5.1    1972 San Francisco Against 2003 Database

Puckett can be innocent if he got a false identification from the database. Although it may sound similar to what Puckett's defense argued, the actual probability is different. The difference lies in that not every member of the database who would obtain a falsely positive identification would be a viable suspect. Time and place already offer strong filters. The scan of the 2003 database may have pointed to someone who in 1972 was not yet born or was a small child or to someone who could not have been in San Francisco at that time, perhaps due to incarceration, foreign military service, or other solid alibi. Gender has the same consequence. A positive identification of a female member of the database would not lead to a plausible accusation.

The possibility of these alternative identifications that would not lead to a prosecution means that not every false positive from the database could correspond to Puckett. Plausible would only be an identification of a male member, of the appropriate age, who may have been in San Francisco in 1972.

In other words, we seek to estimate the intersection of two sets, the population of 1972 San Francisco and the 2003 database. Recognizing this, means that this number can also be approached from the population of the city. The above narrowings of the database tried to get to the size of that intersection from the database, but starting from the population of the city in 1972, one can analogously ask how many of the city's male residents would be in the database. Either path leads to an estimate of the population of the intersection.

Figure 8 illustrates the notion of the overlap, the intersection of the populations of the database and the city. To make the illustration tractable, the city's population is stylized to one thousand and that of the database to two thousand. Estimating the size of the intersection is necessary to form an expectation about the probability of a false positive. The intersection also reveals the error of the argument of the defense that the probability of a false positive comes from raising the accuracy of the test to the power of a number equal to the population of the

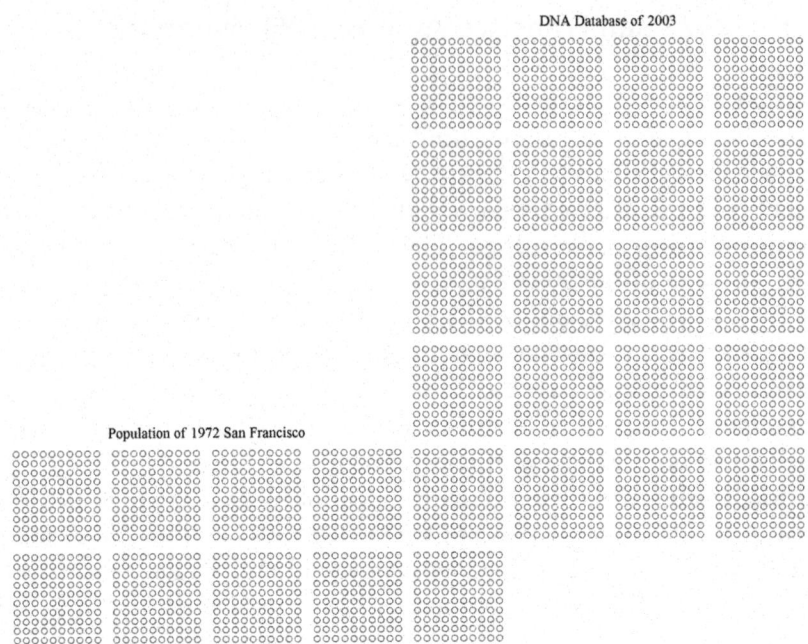

**Fig. 8** Population and database

database. The database likely contains numerous individuals who would not be viable suspects.

Figure 9 illustrates the process of trying to estimate the size of the intersection, the shared population of the city and the database. If one starts from the database, the estimate comes as a result of estimating successively smaller subsets of the database, such as its percentage male, the percentage of the males who are Caucasian, the percentage of those who have the appropriate age, and the percentage of those who do not have an alibi. If one starts from the population of the city, the sequence of subsets may be percentage male, percentage Caucasian, and percentage who end up in the database.

The preceding discussion explained how to move from the raw number of the population in the database to estimating the intersection of the database with the population of potential suspects. The size of the intersection becomes necessary in the next steps. The next steps rely on

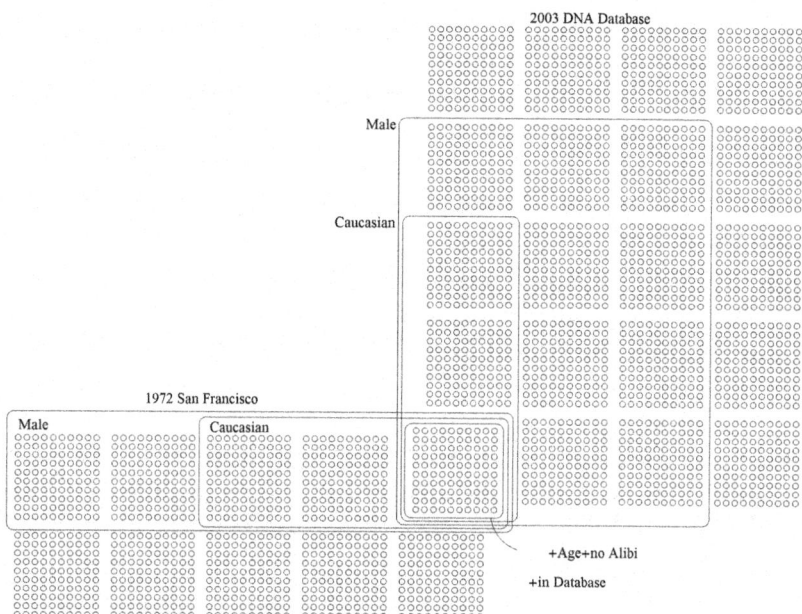

**Fig. 9** Diminishing sets to estimate intersection

realizing that, having received one positive DNA test, we are in one of three alternative states, or, more accurately, in a state produced from one of three alternative sequences of events.

## 5.2    The Alternative Suspect Did It

The first alternative is that the suspect that was investigated but not charged at the time of the crime is the true perpetrator. The reason that the government did not proceed to prosecute him despite that he was the true perpetrator is not relevant, although the probability of that will influence the assessment of how likely this alternative is. If the alternative suspect is the true perpetrator, then the DNA identification is a false positive. Conditional on the alternative suspect being the perpetrator, the probability of one false positive is an application of the binomial distribution.

The binomial distribution, which we also saw applied in the discussion of coin tosses, pp. 56–60, gives the probability of a specific number of successes given the probability of success of a random trial (such as a coin toss or an imperfect DNA test) and the number of times the trial is applied. Here, the binomial distribution can be applied either using the accuracy of the test evaluated for one failure or the inaccuracy of the test evaluated for one success. Thus, the binomial distribution uses as the number of trials the number, $n$, of individuals in the intersection of the population and the database as estimated above. The accuracy of the test in rejecting false identifications, $r$, is the probability of a success if the number of successes is set to $n - 1$. Otherwise the probability of success is $1 - r$ and the number of successes is $1$.[3]

Figure 10 presents this state in a close up of the previous figures that focuses on the intersection of the database and the population. A random member of the population that is not in the database is the alternative suspect, marked A. One random member of the intersection of the database and the population produces a positive identification, the dark point. That is Puckett according to this chain of events. This figure offers a stylized setting for assisting in the visualization. The estimate of the intersection of the population and the database holds one hundred individuals, much fewer than an actual estimate of the intersection of the population of San Francisco and the database. As one false positive appears in the intersection of a hundred individuals, the implied accuracy of the test for a 1% probability of a false positive would be

---

[3]To test this equivalence in Mathematica, we can use the triple equal sign (which is the abbreviation of the equality test) on the appropriate part of the output of the probability distribution function of the binomial distribution. For a number of trials $n$ (the size of the intersection of the database and the population) and DNA test accuracy in rejecting false identifications $r$, the code is:

```
bd1=PDF[BinomialDistribution[n,r],n-1]
bd2=PDF[BinomialDistribution[n,1-r],1]
bd1[[1,1,1]]===bd2[[1,1,1]]
```

which gives $n(1 - r)r^{n-1}$ for both cases and reports that the comparison is true. The intuition behind this formula is that obtaining an outcome with one failure comes from $n - 1$ successes, which have probability $r^{n-1}$, and one failure, which has probability $r - 1$. Multiplying those is not enough, however, because many ways exist to obtain only one positive. The first subject may produce it, the second, the third, or the last. Therefore, we multiply the product by that number, $n$, and obtain the above formula, $n(1 - r)r^{n-1}$.

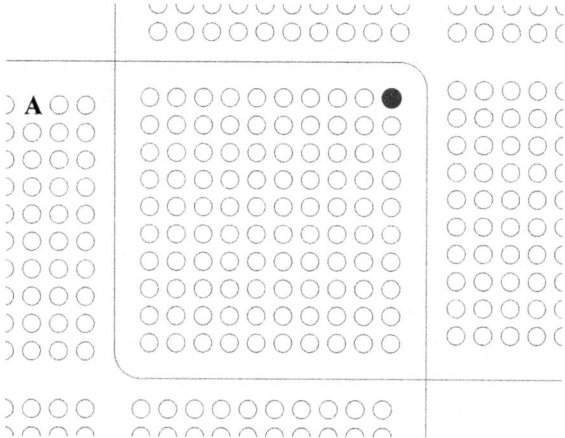

**Fig. 10** The perpetrator is an alternative suspect who was not prosecuted

about 94%, again much lower than the actual DNA test in the case and vastly lower than DNA tests that do not have its problems of aged and degraded samples.[4]

A valid question arises on why this chain of events is separate from the second chain of events, where the perpetrator is an unknown person who is not in the database. The reason for considering separately the suspect who was investigated relates to the reasons that made the police investigate this suspect, Baker. In some sense, something made the police at that time suspect that Baker might have committed the crime rather than anyone else in San Francisco. Thus, Baker is unlike the remaining population of San Francisco. Some additional probability weight falls on Baker being the perpetrator. The fact-finder will need to assess this probability. If the fact-finder believes that the government would never abandon the prosecution of a true perpetrator, then that estimate would be zero. If, at the opposite extreme, the fact-finder believes that the government routinely fails to prosecute true

---

[4]Obtain the implied accuracy of the test by solving numerically the binomial distribution as seen in the previous footnote with

```
FindRoot[(bd1[[1,1,1]]/.n->100)==.01,{r,.95}]
```

which answers .93696.

perpetrators under these circumstances, then the probability of Baker being the true perpetrator would be high. The stylized setting of these three figures, the three alternative chains of events, is apt if the fact-finder treats the probability of Baker being the perpetrator as one-third. In other words, the visualization rests on the idea that the three alternatives are equally likely.

## 5.3    Someone Not in the Database Did It

The second alternative is that the perpetrator is someone not in the data-base, an unknown member of the population of San Francisco. The chain of events is very similar to the prior one. Figure 11 illustrates this alter-native by marking the unknown perpetrator with a U. The perpetrator is located outside the intersection of the population of San Francisco and the database. Again, one point in the database receives a positive DNA identification, which is necessarily a false positive. That is Puckett. Again, the probability of one false positive comes from the binomial distribution using the same analysis as in the case where the perpetrator was Baker.

Again, the fact-finder must establish an estimate of how likely it is that the perpetrator did not end up in the database. The opposite

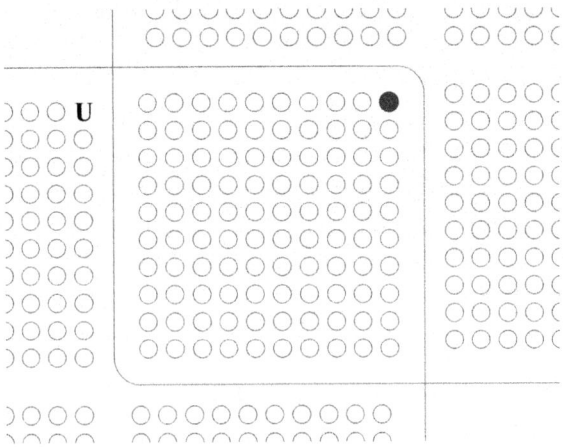

**Fig. 11**  Unknown perpetrator not in the database

extremes here are an extreme confidence in recidivism and police effectiveness versus a view of the crime as a random one-time event that anyone could commit. A fact-finder who believes in the high recidivism of criminals and the effectiveness of police would be confident that the perpetrator of this crime would continue to commit crimes. The over two decades of intervening time imply several crimes and confidence in police and prosecutorial effectiveness means that such a perpetrator would tend to be in the database with very high probability. This would leave a very small probability of the perpetrator being outside it and a small probability for this chain of events having materialized. If, on the opposite extreme, the fact-finder considers this crime as something that anyone might have committed as a one-time event given the circumstances, then the fact-finder would assign large probability to the perpetrator being outside the database. For the three figures to be considered equally likely, after a one-third probability is taken by the first alternative, that Baker was the perpetrator, the fact-finder should assign a probability of fifty percent to the perpetrator being outside the database. An estimate based on extrapolating from actual recidivism and apprehensions places the probability of being outside the database at about 25%.[5]

## 5.4    Someone in the Database Did It

The last alternative chain of events is that someone in the database committed the crime. Then, with the probability of the accuracy of the test, the perpetrator received a true positive DNA identification. Further, however, no additional positives arise because only one positive exists. A second alternative is that the perpetrator might receive a false negative. This flows from the accuracy of the test. The probability of a false negative is one minus the accuracy of the test. Then, if a false negative does occur, then a small probability of a subsequent false positive exists.

---

[5] *See generally* Ian Ayres & Barry Nalebuff, *The Rule of Probabilities: A Practical Approach for Applying Bayes' Rule to the Analysis of DNA Evidence*, 67 Stan. L. Rev. 1447 (2015).

Thus, the binomial distribution needs to give two items. First, the probability that after the perpetrator receives a true positive, no additional positives appear. Second, the probability that after the perpetrator receives a false negative, one (false) positive appears.

The inputs into the binomial distribution are three, the probability of a success, the number of trials, and the number of successes. In both cases, the relevant probability is the accuracy of the test in rejecting false identifications, $r$. The number of trials in both cases is one less than the number of individuals in the intersection, because one is removed by virtue of being the perpetrator who has already either received the positive or the negative. In the first alternative, where no additional positives appear, the number of successful rejections must be all the innocent individuals to whom the test is applied, i.e., $n - 1$. Since all must be successes, the binomial distribution reverts to the easy case of its extreme. The probability that all applications of the test obtain correctly negative results is the probability that the test correctly rejects multiplied by itself, i.e., raised to the power of the number individuals, i.e., $r^{n-1}$. In the second alternative, the number of successful rejections must be one fewer than the number of persons in the intersection in order to have one failure that produces the false positive. The number of successful rejections, therefore, must be $n - 2$. With that adjustment, the formula reverts to what we saw in the first two alternative states.[6]

Figure 12 offers a way to visualize the setting where the perpetrator is in the database. Instead of the intersection holding a hundred individuals, their probability weight unites into a single dark point. The overwhelming likelihood if the perpetrator is in the database is that the perpetrator will receive a true positive DNA identification. Yet, not the entire point is dark. The point is missing a sliver that corresponds to the probability that the perpetrator will receive a false negative. A fraction of that sliver, however, is dark, corresponding to a false positive after the false negative.

---

[6]To whit,

```
PDF[BinomialDistribution[n-1,r],n-2]
```

which produces $(n - 1)(1 - r)r^{n-2}$. Again, the intuition is that the $n - 2$ successful rejections have probability $r$ raised to that power, which is the last term of the formula. The probability of a false positive is $1 - r$, and the product of these two must be multiplied by the number of ways to obtain one positive which is equal to the number of trials, $n - 1$.

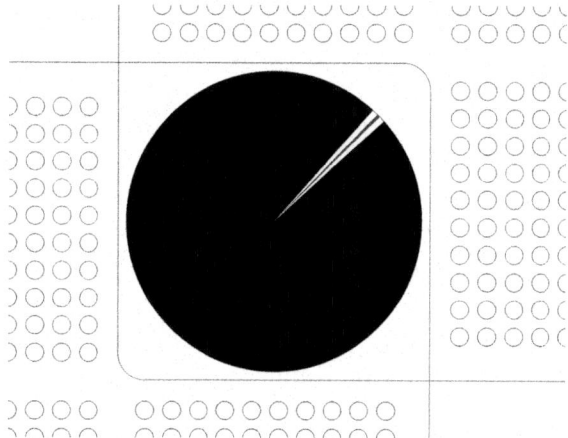

**Fig. 12**  Perpetrator is in the database

These three figures, without further refining the probabilities, can set a framework for assessing the probability of Puckett's guilt. The predicament of the legal system is having observed a positive, having landed on a dark point, but not knowing if that is a true positive, black in the figures, or a false positive, dark gray in the figures. The black is about 98%, the final large point minus the sliver. The dark gray is two percent, the first two graphs, one percent each, plus the fraction of the sliver in the third. Even using the crude graphical approach suggests that Puckett's probability of guilt is quite high. A more exact calculation springs from refining the probability tree by assigning to each branch the appropriate probability.

## 5.5  Completing the Analysis

The above analysis revealed that it includes three values of probability that are a matter of estimation rather than DNA science. The first is the probability $p_a$ that the alternative suspect, Baker, committed the crime. The second is the probability $p_d$ that an unknown perpetrator is in the database. The fact-finder will also have to estimate the size $n$ of the intersection of the database and the population. Nevertheless, the impact of those variables is not large in the final calculation.

The remaining probabilities, which have a large impact on the calculation, are functions of the accuracy $v$ with which a test produces a correct positive identification and the probability $r$ with which the test correctly rejects a negative identification. The accuracy in recognizing positives is a matter of administration of DNA laboratories in the sense that a false negative can only arise from an erroneous reading of the DNA sample due to contamination or a similar error. The record of DNA labs reveals two instances of such an error that led to improved and more secure processes.[7] The accuracy of negatives depends on the distribution of DNA markers in the population and is a matter of DNA science.[8]

Figure 13 reproduces the probability tree of Fig. 7 with the addition along each branch of the corresponding probability. As we saw in the simpler instances of the rare disease test or the cab color witness, the objective is to sum the probability of all the states in which the observed signal appears. Those are the italicized ends of the probability tree. The sum of those probabilities forms the denominator, the total probability mass of observing a single positive, the total weight of the dark points of the figures. The numerator is the true positive, the italicized end of the probability tree that is also bolded. The probability of each is the product of multiplying the probabilities of the branches leading to each end.

A set of estimates gives an example of the probability of guilt in this setting. Set the probability that the alternative investigated suspect was the perpetrator at 20% ($p_a = .2$), set the probability that the perpetrator ended up in the database at 60% ($p_d = .6$), let the probability of

---

[7]The National Research Council explains that it cannot propose such a probability of error:

> There has been much publicity about ... errors made by Cellmark in 1988 and 1989, the first years of its operation. Two matching errors were made in comparing 125 test samples, for an error rate of 1.6% in that batch. The causes of the two errors were discovered, and sample-handling procedures were modified to prevent their recurrence. There have been no errors in 450 additional tests through 1994. Clearly, an estimate of 0.35% (2/575) is inappropriate[ly high] as a measure of the chance of error at Cellmark today.

THE NATIONAL RESEARCH COUNCIL, THE EVALUATION OF FORENSIC DNA EVIDENCE 86 (1996).

Rather, the implied error rate should be much smaller, especially assuming the recommended safeguards that include repeat testing by different laboratories.

[8]*Id.*

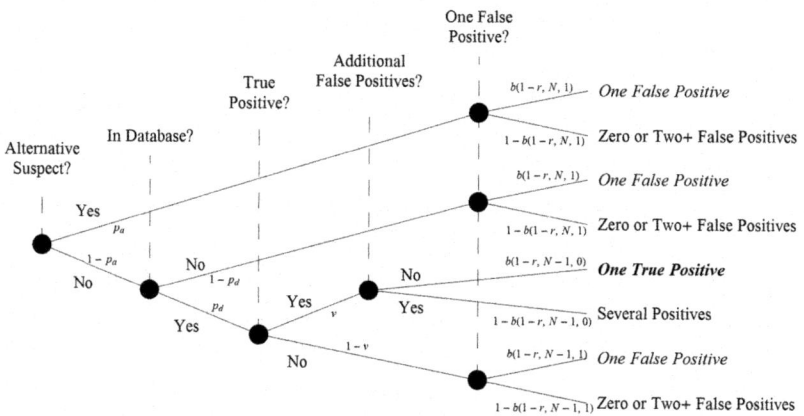

**Fig. 13**   Puckett's probability tree with symbolic values

a true positive be the same as a false negative and equal to the prosecution's testimony about the test's accuracy of one error in 1.1 million $(v = r = (1.1\text{M} - 1)/1.1\text{M})$, and let the intersection of the population and the database $n$ be 8790. Then, the probability of guilt is 99.14%. A spreadsheet or a Mathematica cell with sliders setting the variables is easy to construct, letting users choose their own parameters.

At its core, this example may also hide important lessons about the role of the jury as the trier of fact in the Anglo-American criminal trial. Establishing a value of a probability of guilt may well be something very different than deciding that an accused is guilty. A jury may convict on tenuous testimony of police officers but acquit on the probabilistic nature of the proof of guilt perhaps on reasons with little relation to actual probability of guilt.[9]

---

[9]*See, e.g.*, David N. Dorfman, *Proving the Lie: Litigating Police Credibility*, 26 Am. J. Crim. L. 455 (1999); Christopher Slobogin, *Testilying: Police Perjury and What to Do About It*, 67 U. Colo. L. Rev. 1037, n. 14 (1996); Myron W. Orfield, Jr., *The Exclusionary Rule and Deterrence: An Empirical Study of Chicago Narcotics Officers*, 54 U. Chi. L. Rev. 1016 (1987). Alan Dershowitz points out:

I have seen trial judges pretend to believe officers whose testimony is contradicted by common sense, documentary evidence and even unambiguous tape recordings ... Some judges refuse to close their eyes to perjury, but they are the rare exception to the rule of blindness, deafness and muteness that guides the vast majority of judges and prosecutors.

Alan N. Dershowitz, *Controlling the Cops; Accomplices to Perjury*, N.Y. Times, May 2, 1994, at A17.

# 6    Judicial and Prosecutorial Error

The advent of DNA evidence also had a salutary effect for many defendants. Scores of falsely convicted have been released through the activities of organizations such as the Innocence Project. The result is that for the last few years before the development of DNA identification technology, a natural experiment arose. Prosecution and conviction occurred without the benefit of DNA identification. Then, DNA identification tested the accuracy of the outcomes of the legal system. The result is that we can find the rate at which criminal trials convicted innocent defendants. Granted, the process still requires some estimation. Nevertheless, scholars have proposed a narrow band as their estimate of the accuracy of the criminal system.

To recognize the problem, start with a simplified setting of one hundred observed convictions, which subsequently receive DNA tests, and assume the tests are accurate. The DNA tests indicate that five of the hundred are innocent. The appearance is that the error rate was 5%. Yet, this may be an oversimplification. Suppose that the 95 convicted, who were not exonerated by DNA, had trials in which multiple eyewitnesses testified against the accused. By contrast, the five falsely convicted were all convicted on circumstantial evidence or plea bargains. Under these assumptions, the accuracy of trials with circumstantial evidence and that of plea bargains is zero, whereas the accuracy of trials with multiple eyewitnesses is perfect. In other words, the problem for estimating the accuracy of the criminal justice system is that not all convictions are alike. Convictions on circumstantial evidence are not rare. Plea bargains, rather than being rare, are the predominant path to convictions.[10]

Despite the hurdles, Professor D. Michael Risinger tries to address the accuracy of trials by focusing only on capital rape-murders, meaning murders accompanied by rape where the capital punishment was

---

[10]Estimates from 2004 place the proportion of plea bargains at over 97% of federal convictions and a similar estimate for state convictions. *See* Jed S. Rakoff, *Why Innocent People Plead Guilty*, New York Review of Books (Nov. 20, 2014), http://www.nybooks.com/articles/2014/11/20/why-innocent-people-plead-guilty/ [perma.cc/ZBU3-JAFF.com.F].

imposed. In such crimes, eyewitnesses are relatively rare, and the imposition of the capital punishment mostly precludes plea bargains.[11] The lowest estimate Professor Risinger produces is 3.3%, i.e., he estimates that at least 3.3% of such trials produce a false conviction. The error rate could be higher, depending on what additional trials should be appropriately excluded as not being truly comparable to the circumstances in which the DNA exonerations arose. Consider, for example, that multiple eyewitness testimony never leads to a false conviction. Consider also that all trials where DNA evidence did not exonerate the accused, had multiple eyewitness testimony. Then, the appropriate sample of trials in which DNA exonerations arose in capital rapes would be those trials that did not have multiple eyewitnesses. In that subsample of trials, the exonerations would be 100%. Professor Risinger's estimate of the number of trials where the circumstances leave little doubt about guilt places the estimate of the maximum error rate at 5%.

The point here is not to review the analysis of Professor Risinger, which is thorough and persuasive. Rather, let us explore a complication that is by nature excluded in the analysis, that of plea bargains. Generally speaking, the idea behind plea bargains is that the prosecutor accepts a guilty plea to a lesser crime rather than go through trial and obtain a conviction to a greater crime but which would be subject to the uncertainties of trial. For the accused, the plea bargain similarly offers a conviction to a lesser crime, avoiding the risk of conviction after trial to a greater crime.

Returning to the question of the accuracy of the criminal system, the question arises how the accuracy of plea bargains interacts with the accuracy of criminal trials to reach the overall accuracy of the criminal justice system. Perhaps plea bargains reduce overall accuracy but the possibility exists that prosecutors who offer plea bargains, unburdened by the procedural limitations of trials, may be more accurate than trials, and plea bargains may improve overall accuracy. Before even getting to this point, the problem is merely how to model the resulting successive filters of plea bargains and trials leading to the resulting overall accuracy.

---

[11]D. Michael Risinger, *Innocents Convicted: An Empirically Justified Factual Wrongful Conviction Rate*, 97 J. CRIM. L. & CRIM. 761–806 (2007).

One approach to modelling judicial and prosecutorial error starts with accuracy of trials. Akin to the examples of the disease or the cabs, the population consists of two classes, the innocent and the guilty. Trials produce signals of guilt or innocence with some accuracy. The result is four groups: (a) innocent who receive an accurate verdict of innocent; (b) innocent who receive a false verdict of guilty; (c) guilty who receive an accurate verdict of guilty; and (d) guilty who receive a false verdict of innocent. This, essentially, replicates the settings of the disease or the cab example. The effect of plea bargains becomes visible by adding before trial the stage of plea bargaining. The accused can accept a plea of guilty or go to trial. Just as both innocent and guilty go to trial, both innocent and guilty receive plea bargains. Each group accepts the plea bargain with some probability. However, this approach misses the accuracy of prosecutions. To observe the accuracy of prosecutions their filter must appear in both paths. In the path with the plea bargain, society has the innocent and the guilty with some proportion, they face prosecution with some accuracy, accept a plea bargain with a third rate, and those who do not plea undergo trials that have a fourth rate of accuracy. Without plea bargains, simply omit that stage.

From the perspective of this model, the 3.5% rate of false convictions that Risinger observes is the result of both the accuracy of prosecutions and trials. It can be a result of various combinations of accuracies at the two stages. Perhaps an interesting approach would be to calculate the pairs of accuracies that produce that result. Placing the accuracy of trials on the horizontal axis and the accuracy of prosecutions on the vertical axis, what combinations of the two produce convicted populations containing innocents at the rates of, say, 4 and 10%?

Begin by positing that the population commits crimes corresponding to a rate of guilty persons in the population $g$. The prosecutor has accuracy $r$ and trials have accuracy $t$. The setting omits plea bargains for now and is similar to the examples of the cab and the disease paradoxes but using two inaccurate signals instead of a single one. The population divides into several groups: the guilty and the innocent who may be prosecuted or not, and, if prosecuted, convicted or not.

The guilty are prosecuted with a rate $r$, giving $gr$ as the fraction of the population that is guilty and prosecuted, with $g(1 - r)$ being the guilty who are not prosecuted. Those guilty who are prosecuted also are convicted with rate $t$, giving $grt$ as the fraction of the population that is guilty and convicted, with $gr(1-t)$ being the guilty who are prosecuted but not convicted.

The innocent are prosecuted with rate $(1-r)$, giving $(1-g)(1-r)$ as the fraction of the population that is innocent and prosecuted, with $(1-g)r$ being the innocent who are not prosecuted. Those innocent who are prosecuted also are convicted with rate $(1-t)$, giving $(1-g)(1-r)(1-t)$ as the fraction of the population that is innocent and convicted, with $(1-g)(1-r)t$ being the innocent who are prosecuted but not convicted. Verify that all population groups are identified by ensuring that the sum is 1.

The composition of the convicted is the guilty convicted plus the innocent convicted, namely $grt + (1-g)(1-r)(1-t)$. That becomes the denominator of the fraction showing the innocent as a fraction of the convicted population. The second term of the sum becomes the numerator of the corresponding fraction. Solve for the innocent being a given fraction of the convicted population. Plot for the innocent being 4 and 10% of the convicted population with the guilty as a percentage of the total population taking the values of 2, 1, and .01%. The result is Fig. 14 (but its production is left as an exercise, see Exercise 9).

The three figures correspond to three different proportions of the population being guilty. The left graph corresponds to 2% of the population being guilty, the middle one to 1%, and the one on the right to .01%. Each graph shows two lines, corresponding to two values of the convicted population that is innocent, the solid line corresponds to

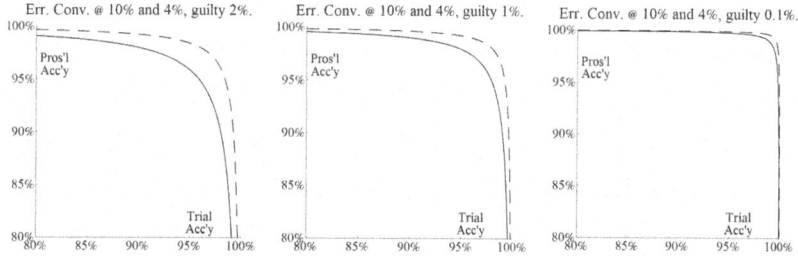

**Fig. 14** Prosecutorial and trial accuracies

10% and the dashing line to 4%. The general pattern is that inaccuracy by either the prosecutor or the court means that the other has to be very accurate for such rates of false incarceration to exist. The intensity of this effect is most pronounced in the case of the fewest guilty, .01% of the population, the figure on the right, where any accuracy below 99% by one of the two requires the other to have nearly 100% accuracy. The least extreme case, where the guilty are 2% of the population, produces the falsely incarcerated values in question (10% and 4%) with slightly lower accuracies of prosecution and trial, so that when each has an accuracy of about 96%, then the rate of the falsely incarcerated is 10%.

Consider the error rate observed in capital rapes of about 3.5%. What accuracies of prosecutors and courts produce them? How do those compare to our everyday experience with errors? My reaction is that the accuracies needed to produce this low a false incarceration rate are quite high but saddled with the inescapable human fallibility.

# 7    Misperception of Probability: The Three Boxes Problem

Legendary is the three boxes problem, also known as the Monty Hall problem from the name of the announcer of a television game that used it. For those who do not know the problem, it is confounding. In part, the point of examining it to understand a curious application of probability theory. Mostly, however, consider it a cautionary note: that the human brain often misperceives probabilistic challenges.

The player faces three closed boxes. Two boxes are empty; one box contains a prize. The player selects a box at random, hoping to pick the one containing the prize. The host, who knows where the prize is, opens one of the boxes that the player did not pick, and reveals that it is empty. The host offers to the player the opportunity to switch to the remaining closed box or to stick with the player's original selection. What should the player do?

When one encounters this problem for the first time, it seems that the player's choice does not matter. The player started with a one-in-three chance of having selected the box with the prize and it seems that switching simply moves the choice to a different one-of-three box, with no change in the probability of getting the prize. This is wrong, however. Switching gives a two-thirds chance of winning while sticking with the player's original choice gives the player a one-third chance of winning. Explaining the error is extremely difficult.

One approach to perceiving the error is to suppose that the game has a hundred boxes. The player selects one. Focus, though, on the probability that the prize is in one of the other boxes. That is 99%, undoubtedly. Now, the host opens 98 of the other boxes so that only two boxes are left closed, the one the player picked and one more. The chance that the prize is in the player's box has not changed, it is still 1%. The chance that the prize is not in the player's box has also not changed, it is 99%. Switching gives a 99% chance of winning. Analogously, in the three boxes setting, switching produces a two-thirds chance of winning.

One more approach to understanding the three boxes problem is to focus on the host's knowledge of which boxes are empty. The host's opening of a box is, thus, predetermined, it does not contain information. The true choice is for the player to stick with the original choice of the box and its small probability of success, or to switch to all the other boxes and win if the prize is in any one of them.

Table 2 offers a slightly more analytical approach to the result of the two strategies, sticking or switching. Call the box that the player selects the picked box, and the other two boxes the next and the last. The first column shows the three boxes. The prize can be in one of the three boxes, the picked, the next, or the last, with equal probability of 1/3, in the second column. The third column shows the box that the host opens. The last two columns hold the two strategies, sticking with the originally selected box, in column 4, or switching to the remaining closed box, which is column 5. Return to the third column, the host's action. When the prize is in the picked box, the host opens either the

**Table 2**  The three boxes strategies

| Prize in | Prob. | Host opens | Sticking | Switching |
|----------|-------|------------|----------|-----------|
| Picked | 1/3 | Next or last | Wins | Loses |
| Next | 1/3 | Last | Loses | Wins |
| Last | 1/3 | Next | Loses | Wins |

next or the last box, in which case sticking wins (column 4) and switching loses (column 5). When the prize is in either the next or the last box, the last two rows of the table, then the host opens the other one and switching wins, two-thirds of the time. The last two columns, corresponding to the two strategies, sticking or switching, show that sticking with the picked box wins one-third of the time whereas switching wins two-thirds of the time.

No amount of explaining makes the three boxes problem easy to understand. It is one of the best illustrations that our minds seem to have a structural difficulty in comprehending some probabilistic settings.

# 8     Inference, Confidence, Tests

Of all the probability theory topics that, perforce, cannot receive appropriate coverage in such a swift overview, the inference (also called estimation) of properties from samples is probably the most important. Sampling is the foundation for statistical inference. Often, we only observe samples of the populations that random processes generate. What inferences can we draw about the characteristics of the entire populations?

In the setting of a single sample, this question changes to what is the best estimate for the mean and standard deviation of the population. It turns out that the best estimate of the arithmetic mean is the observed mean but the best estimate of standard deviation requires an adjustment. Instead of dividing the square root of the squared differences from the mean by the number of the observations, the best estimate of the standard deviation of the entire population requires that we divide by one less than the observations. In formal notation, after summing the squares of the difference of each observation, $m$, from the mean, $M$, divide by $N - 1$ before taking the square root. Namely, the estimate of the standard deviation $s_s$ derived from a sample is

$$s_s = \sqrt{\frac{\sum (m - M)^2}{N - 1}}.$$

Accordingly, when estimating outcomes from a generative process that is sampled rather than truly known, the estimated sample standard deviation applies. This is also called Bessel's correction. While the full explanation of why the adjustment is to divide by $N - 1$ is technical, an intuition may be this. We seek to estimate the true dispersion of the population from a sample and dispersion is a measure of the differences between the elements of the population. A sample of one element gives no information about dispersion but does give information about the mean. If we add a second element to our sample, then we only experience one difference despite that the calculation of the mean makes it into two. The one difference we see is our best estimate of difference in the sample. Adding a third element, only adds to our experience one additional element of difference (from the mean of the prior two elements), and so on. We care about differences, not elements, and the differences are one fewer than the elements of the sample.

This (sample) standard deviation determines the confidence of the resulting estimated mean. Often researchers will report a 95% confidence interval. This means they will subtract two (sample) standard deviations from the estimated mean and use that as the lower value of the confidence interval, then add them and use that value as the upper bound of the confidence interval. Given the data, with 95% probability the true mean lies inside this confidence interval. Figure 15 illustrates how over 95% of the probability weight lies within two standard deviations. The figure uses the standard normal distribution, i.e., the normal distribution with a mean of zero and a standard deviation of one. Using the normal distribution is appropriate because all random processes, if they comprise many other random processes, tend to approach the

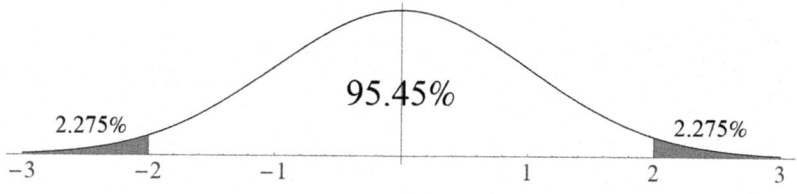

**Fig. 15**  The probability of being within two standard deviations

normal distribution according to a law of probability theory called the central limit theorem (think of the difference of how the outcomes of a single coin toss are distributed to the outcomes of the sum of a hundred coin tosses; the latter approaches the normal bell curve, see Fig. 2 in Chapter 5, p. 58). The probability weight beyond two standard deviations, marked in gray in the figure, is very small, 2.275%. The probability of being within two standard deviations of the mean is 95.45%. For readers who understand the role of cumulative distribution functions and probability density functions of distributions (this bell curve is the probability density function of the standard normal distribution), Exercise 6 asks the derivation of the values and the graph of Fig. 15.

Estimation from samples takes additional complexity when dealing with more complex problems, for example, when comparing a sample to a null hypothesis or when comparing two samples. The null hypothesis is the counterfactual that the researcher's hypothesis is false, for example, that a treatment has no effect. To use one of the examples discussed, consider the paper on the relation between abortion and crime 18 years later (see Fig. 1 in Chapter 7 and accompanying text, p. 98). The research finds that the legalization of abortion reduces crime. The corresponding null hypothesis is that the legalization of abortion did not reduce crime.

The pertinent question for statistical analysis is what is the probability of observing the sample if the null hypothesis is true. In visual terms, the bell curve is placed at the null hypothesis rather than where the data are. Although the null hypothesis drives the location of the bell curve, the standard deviation of the bell curve depends on the data, i.e., the data determine how fat its tails will be. In the case of two samples, the usual question is what is the probability that the two samples were produced by the same random process. In visual terms, a single bell curve is placed as if both sets of data were one and then the statistical test calculates the probability of observing the data if the null hypothesis were true.

Using the example of the relation of crime to abortion, the statistical test answers the question what is the probability of observing this little crime if crime after the legalization of abortion were produced by the same random process that existed before the legalization. As the probability of observing the difference that the data reveal drops, the researcher gains confidence in saying that the statistical analysis rejects the null hypothesis with the corresponding level of confidence. If, for example, the data could only arise by chance with 1% probability if the null hypothesis were true, then the researcher states that the data reject the null hypothesis with 99% confidence.

The statistical tests depend on differences. The resulting methods of analysis for observations that fall into categories are the chi-test or chi-squared test (more accurately, the $\chi^2$ test), and for differences of an estimate from a null hypothesis or from a different sample, the $t$-test. Those rely on derivations of distributions (the $\chi$-distribution and the $t$-distribution) that the corresponding metrics would have, depending on the size of the samples (again, approaching the normal distribution as the sample size increases). These two tests are included in most quantitative and spreadsheet software, making them quite easy to deploy.

# 9    Conclusion

The world of finance and business, as well as life generally, is full of uncertainties. Frustratingly, their proper analysis is often quite counterintuitive. To understand them correctly, one needs to multiply the setting and explore hypothetical settings of entire populations. In the rare disease paradox, the answer to the probability of the test being accurate lies in visualizing the entire population being tested, and calculating true and false positives. In the cab color paradox, the answer to the accuracy of that cab's color lies in supposing that the witness calls the color of every cab in a multitude. Passing to the probability of the accuracy of a DNA test pursuant to a sweep through a database, one must envision alternative worlds of suspects, and false and accurate tests. The three boxes problem defies intuition, but again the key is exploring all possible outcomes. After undermining our confidence in our reasoning, the chapter closes with a reference to conventional statistical estimation, confidence intervals, and tests of hypotheses.

## 10    Exercises

1. You are going on a trip to two cities that are sufficiently far apart that their weather is not related (i.e., the risk of rain is independent). The weather forecast is for a 20% chance of rain in the first city and a 25% chance of rain in the second city. What is the probability that you experience no rain in either city? What is the probability that you experience rain in both cities?

2. You are going on a trip to two cities that are not far from each other so that they tend to be subject to the same weather phenomena. The weather forecast is for a 20% chance of rain in the first city and a 25% chance of rain in the second city. (They can have different probabilities of rain because of the path and size of the weather.) What is the probability that you experience no rain in either city? What is the probability that you experience rain in both cities?

3. A client is considering a long-term investment in a region hit by a devastating natural phenomenon, which occurs with 3% probability each year (think of hurricanes, mudslides, earthquakes, etc.). Over a period of ten years, what is the probability that the region will suffer at least one such phenomenon? Over a span of ten years, what is the probability that the region will suffer exactly one such phenomenon? Hint: The event may occur in the first year and not occur in any other year, or the event might occur in the second year and not occur in any other year, and so on. (Compare the discussion of obtaining one false positive identification in the various stages of the analysis of the DNA problem, *see* text accompanying Footnote 3.)

4. An insurance company insures a portfolio of homes in two cities subject to rare but severe natural phenomena. The cities are sufficiently far apart that their risk is independent. The first city has a 2% probability of experiencing such a phenomenon annually and the second city has a 1% probability. What is the probability that homes insured by the insurance company will experience one or more disasters over ten years? What is the probability that in a ten-year span, both cities experience a natural disaster in the same year (once or in more than

one year)? What is the probability that both cities experience a natural disaster in the same year exactly once during the ten years?

5. Imagine an alternative way that the personal computer industry unfolded. Imagine that at some point in the early 1980s, three competitors existed: IBM, Apple, and Microsoft. No others could credibly threaten to enter the industry, which was still in its nascent stage. Each firm defends its position with several key patents. Due to the network effects of software, one of the three firms will eventually dominate the industry, producing enormous gains for its shareholders, leaving the other two firms with little value. Some investors and clients of yours make a large investment in one of the three, say Apple. Then, in response to a patent challenge, the courts find that all the patents of a different one of the three, say IBM, violate the patents of the other two. The investors receive an opportunity to switch their investment to the remaining company, Microsoft. Discuss the similarities and differences of this setting from the three boxes problem.

6. Distributions are determined by their cumulative distribution functions (CDFs), which describe the probability of an outcome smaller than their parameter. CDFs take two parameters as input, the distribution and the value in question. The values that are the output of CDFs go from zero, at the minimum of the distribution in question, to one, at the maximum. The bell curve is the probability density function (PDF), which is the derivative of the CDF. Its inputs are also the distribution and the value where it is to be evaluated. Use the CDF and PDF of the normal distribution (in either Mathematica or Excel), to find the probability of being outside two standard deviations from the mean of a normal distribution and that of being within two standard deviations. Use the PDF to produce Fig. 15.

7. You are concerned that a regular six-sided die has a defect making it produce outcomes of 2 and 3 more often than proper. You have the outcomes of the last 600 casts of this die as producing a one 75 times, a two 132 times, a three 120 times, a four 99 times, a five 80 times, and a six 94 times. Will you apply the chi-test or the $t$-test to determine if the die is defective? A student who has taken statistics

should also be able to deploy the test and answer what is the probability that the die is defective.

8. An orange juice bottler uses two machines to produce bottles of orange juice. The bottler has the weight of each bottle produced the last month. Which test should the bottler use to find if the two machines fill each bottle to the same level?

9. Recreate Fig. 14, p. 141.

# 9

# Financial Statements and Mergers

## 1    The Purpose of Financial Statements

A business can be described from several perspectives. One might wish
to describe the personal characteristics of its individuals or the aesthetics of its premises, but the perspective at issue here is its financial
health. Financial statements aim to describe the affairs of a business in
the units of money that the business uses in its transactions. However,
accountants use financial statements in a fundamentally different way
than lawyers do. Accountants focus on viability, and lawyers focus on
allocation of value. That accountants focus on viability means that
accountants see as a primary threshold problem whether the business
is viable in the sense that the business will make enough money in the
immediate future to pay its expenses. That lawyers focus on allocation means that lawyers envision transactions that exchange the business for other assets, hopefully cash, and ask how that cash should be
allocated among the various claimants, i.e., creditors and stockholders. Creditors can have different seniorities, from secured to ordinary

© The Author(s) 2018                                                                **151**
N. L. Georgakopoulos, *Illustrating Finance Policy with* Mathematica,
Quantitative Perspectives on Behavioral Economics and Finance,
https://doi.org/10.1007/978-3-319-95372-4_9

and subordinated. Stockholders can be of different classes, preferred and common. The stratification of the claimants against a business (the various classes of creditors and stockholders) is called the *capital structure* of the business.

# 2    Balance Sheet and Income Statement

Two are the primary financial statements, one that describes what the business has and what it owes, which is called the *balance sheet*, and one that describes what the business did, which is called the *income statement*. Because accounting terms are not standardized, terms are not always used in an identical way. For example, the income statement may also be called *statement of operations*. Accountants also use a third type of statement, which describes the ways in which the business uses its cash, which is called the *statement of cash flows*, but that will not receive attention here.

The balance sheet consists of two lists of lists, that of *assets* and that of *liabilities and equity*. Conventionally, those two lists appear side by side, so they are also called sides of the balance sheet and, further, the assets can be called the left-hand side of the balance sheet. The liabilities and equity can be called the right-hand side. Accountants focus on the threshold question of whether the business is viable. Therefore, they have designed each side of the balance sheet to have on top the short term. The most liquid assets are on top of the asset side, and the most short-term liabilities are on top on the side of liabilities and equity. This way, one can readily see if the impending expenses are covered by the cash and most liquid assets, such as accounts receivable, i.e., the outstanding bills that the business is about to collect.

This structure of the balance sheet does not help the question of how to allocate the value of a poorly performing business among the different claimants in case of extraordinary events such as liquidation or bankruptcy. In order to determine how to allocate the value of the business, the balance sheet needs to be constructed anew, with the most senior liabilities on top, and if those are secured, matched with the assets that secure them, their *collateral*. Neither the seniority of liabilities nor their collateral relate to the short term or the long term.[1] Listing liabilities by priority, matched with their collateral when secured, allows one to see which claims will not receive full satisfaction if the assets are less than the liabilities. A balance sheet arranged by seniority appears below, see Table 2.

The income statement is also a list of lists, of the money that the business made and the money that it spent, potentially leaving the profit that the business made. To rephrase this in the language of accountants, the income statement is a list of the *revenue* minus *expenses*, leaving the *earnings*. Thus, *revenue* is the amount of money coming into the business in exchange for its products or services. Revenue is also often called *sales*. *Expenses* are the amounts that the business spends, not only to make its product or provide its service, but also including items such as interest, taxes, and some more theoretical amounts such as depreciation and amortization. *Earnings* are the profits made after subtracting expenses from revenue.

The balance sheet aims to display the affairs of a business at a particular moment in time. One can think of the balance sheet as a still picture. The income statement tells a story. One can think of the income statement as the movie that tells how one picture changed into the next one.

The income statement corresponds to an arithmetic operation, that revenue minus expenses equals earnings. The balance sheet also corresponds to an arithmetic operation, that assets equal liabilities plus

---

[1]A weak inverse correlation used to exist. Consider the time before modern security interests in personal property through recordation or filing, i.e., before Article 9 of the Uniform Commercial Code, when security interests in personal property were possessory. Then, such a correlation did exist. Because, to create a security interest in a chattel, one had to deliver possession of the chattel to the lender (the pawnshop), businesses could not use their chattels for secured borrowing. Then, the secured obligations tended to be long-term mortgage loans, and those encumbered real estate. Both the corresponding asset and the liability that it secured, tended to appear at the bottom of each side of the balance sheet. On the asset side, real estate appears at the bottom. On the side of liabilities and equity, long-term loans appear at the bottom of liabilities, however, still above equity, which is junior to all liabilities. So, even then, the liabilities and equity side was not arranged coherently from the perspective of seniority. Now, when with a filing, security interests can be created encumbering even the most liquid assets, such as receivables or cash, any relation between seniority and location on the balance sheet has disappeared.

equity. This is called the *fundamental equation of accounting*, and can be restated as equity equals assets minus liabilities. Here is the source of the term balance sheet. The left-hand side, the assets, is always equal to the right-hand side, the liabilities plus the equity, so it balances.

The fundamental equation of accounting shows that equity has the same derived nature as profits. Equity is not a number that arises independently. Rather it is the result of subtracting liabilities from assets. In an abstract sense, equity is what would be left for the owners of the business if they were able to liquidate its assets for exactly the amount that appears on the balance sheet and if they were able to repay all obligations by spending exactly the amount that appears on the balance sheet. Then, the cash that the owners would receive would be the equity as it appears on the balance sheet. This is a theoretical concept, however, for various reasons.[2] Equity is also sometimes called *net worth* or, especially in banking regulation, *capital*.

Tax issues do not depend on financial accounting. The financial accounting choices regard only what information to present to the investing public. The tax rules, which may be called tax accounting by some, are an entirely different topic; tax and financial accounting only appear related because they share some concepts, like depreciation.

The balance sheet and the income statement are sometimes in tension. What makes for an accurate still picture may lead to poor cinematography and vice versa. One type of tension arises from

---

[2]Neither the assets nor the liabilities would correspond exactly to what appears on the balance sheet. Most importantly, a viable business would be valued as a multiple of its earnings, typically much more than the value of its assets. Even if the business could not be sold as a going concern, assets that lose value over time, like machines or vehicles, do not tend to lose exactly as much value as depreciation removes (moreover, some may remain useful even after they are fully depreciated, in which case they are worth more than the zero to which they would correspond on the balance sheet). Appreciating assets, like real estate, may be worth more than their representation on the balance sheet, which may also be the case with some intellectual property. On the liabilities side, again, some liabilities may have prepayment penalties, making them costlier to repay than what appears on the balance sheet. Long-term liabilities that have fixed interest rate payments and trade in an exchange, like bonds, may produce one more discrepancy. Bonds could be bought instead of being repaid. They could be cheaper to buy than what appears on the balance sheet if interest rates have dropped.

depreciation; there, we will see that the accuracy of the income statement takes precedence over the accuracy of the balance sheet.[3] Turn, now, to an actual balance sheet.

# 3 An Actual Balance Sheet

Before the next section explains the abstraction of the visual balance sheet, let us visit an actual balance sheet. We will observe the chaotic nature of the balance sheet and try to address some of the difficulties in reading it. We visit the balance sheet released on March 4, 2004, by Martha Stewart Living Omnimedia, Inc., ticker symbol MSO. Martha Stewart, the renowned homemaking author and television personality, consolidated her media into MSO, which at its height had a market capitalization of about one billion dollars.[4] The balance sheet appears in Table 1. Keep in mind that this is an auspicious time for MSO, as its head, Martha Stewart, was accused of obstruction of justice in a very public criminal trial about insider trading.

Notice that the numbers that appear in the balance sheet form two columns, one with the heading December 31, 2003, on the left, and one a year earlier on the right. Here appears the first counterintuitive convention of accounting. In most tables displaying data over time, time runs to the right so that the earliest year is on the left and the latest year is on the right. Financial statements are backward, with the latest year on the left.

Observe that some numbers are in parentheses. The parentheses signify that those numbers are negative. This is one more deviation of accounting practice from everyday usage, where a minus sign would indicate a negative value.

Given, perhaps, the type of contemporary communication through the printed page in portrait orientation, notice that the two-sided nature of the balance sheet is lost. The left-hand side—assets—appears first, on top. The right-hand side—liabilities and equity—appears below it.

---

[3] *See* Exercises 5–8.

[4] Market capitalization is the total value of all outstanding shares of a publicly traded business.

**Table 1**    MSO balance sheet released March 4, 2004

Martha Stewart Living Omnimedia, Inc.
Consolidated Balance Sheets
(in thousands, except per share amounts)

|  | December 31, 2003 | December 31, 2002 |
|---|---|---|
| ASSETS | | |
| CURRENT ASSETS | | |
| Cash and cash equivalents | $165,566 | $158,840 |
| Short-term investments | 3,100 | 20,110 |
| Accounts receivable, net | 39,758 | 37,796 |
| Inventories, net | 7,485 | 8,654 |
| Deferred television production costs | 3,465 | 4,179 |
| Deferred income taxes | 10,682 | 7,028 |
| Other current assets | 4,422 | 4,756 |
| Total current assets | 234,478 | 241,363 |
| | | |
| PROPERTY, PLANT, AND EQUIPMENT, net | 22,673 | 31,288 |
| INTANGIBLE ASSETS, net | 44,257 | 44,257 |
| DEFERRED INCOME TAXES | 3,224 | 2,827 |
| OTHER NONCURRENT ASSETS | 4,470 | 4,807 |
| **Total assets** | **$309,102** | **$324,542** |
| | | |
| LIABILITIES AND SHAREHOLDERS' EQUITY | | |
| CURRENT LIABILITIES | | |
| Accounts payable and accrued liabilities | $26,628 | $40,517 |
| Accrued payroll and related costs | 10,360 | 9,385 |
| Income taxes payable | 167 | 323 |
| Current portion of deferred subscription income | 23,833 | 24,932 |
| Total current liabilities | 60,988 | 75,157 |
| | | |
| DEFERRED SUBSCRIPTION INCOME | 7,133 | 7,715 |
| OTHER NONCURRENT LIABILITIES | 4,316 | 5,035 |
| **Total liabilities** | **72,437** | **87,907** |
| | | |
| COMMITMENTS AND CONTINGENCIES | | |
| | | |
| SHAREHOLDERS' EQUITY | | |
| Class A common stock, $0.01 par value, 350,000 shares authorized: 19,628 and 19,342 shares issued in 2003 and 2002, respectively | 196 | 194 |
| Class B common stock, $0.01 par value, 150,000 shares authorized: 30,059 and 30,295 shares outstanding in 2003 and 2002, respectively | 301 | 303 |
| Capital in excess of par value | 183,744 | 181,629 |
| Unamortized restricted stock | (307) | (993) |
| Retained earnings | 53,506 | 56,277 |
| | 237,440 | 237,410 |
| Less class A treasury stock - 59 shares at cost | (775) | (775) |
| **Total shareholders' equity** | **236,665** | **236,635** |
| Total liabilities and shareholders' equity | $309,102 | $324,542 |

Although the balance sheet puts the shortest-term assets and liabilities at the top, accountants do not use the words short term, saying instead *current* and *noncurrent* or *fixed*.

Worth pointing out is the word *consolidated* in the heading of the balance sheet. This is routine in most financial statements and means that this financial statement reflects (consolidates) the affairs of several businesses, likely different subsidiaries and business lines that, strictly speaking, should each have its own balance sheet. For example, perhaps the accounts receivable are strictly speaking not receivable by the parent, MSO, but by various subsidiaries of it. Yet, the consolidated financial statement ignores the separation of those entities and talks about accounts receivable as if the separation did not exist. In contrast to law, where ignoring the separation between entities is highly exceptional and requires the piercing of the corporate veil, accounting statements routinely ignore the separation of affiliated entities by merely using the label *consolidated*.

Immediately under its heading, the balance sheet states that the amounts are in thousands. This makes eminent sense, to avoid all the unnecessary zeros. For example, looking at the second line where short-term investments appear as $3100, the actual value of short-term investments is $3,100,000. The statement provides an exception for the per-share amounts, those are not in thousands, only the amounts that are about the entire corporation are in thousands. Yet, when a per-share amount appears, the $0.01 par value (under shareholders' equity), the accountants assume that the reader knows that par value is stated per share.

Important is a method that hides behind the balance sheet numbers. The values that accountants use to construct the balance sheet are (a) the values of the individual assets rather than the potentially greater value that their combination into businesses may have; and (b) those values usually rest on historical cost rather than current market values. When accountants use the market values in the present for the assets instead of historical cost, then they say that the assets are *marked-to-market* values. These two choices that accounting rules make, produce various distortions when we try to apply the balance sheet to policy questions.

Notice that, in contrast to the original publication, in boldface appear the crucial figures that form the fundamental equation: total assets, total liabilities, and total equity. Whereas one who is learning about the balance sheet would expect them to appear prominently, they are not given any greater prominence in the way that accountants present them. On the contrary, one can argue that they appear ordinary or even secondary.

In essence, the balance sheet tells us that MSO has assets of $309 million, liabilities of $72 million, and equity of $237 million. Before we use those figures for a visual illustration of the balance sheet, let us gain an understanding of depreciation.

# 4    Depreciation

Suppose that MSO uses a machine in its operations that lasts exactly four years. MSO obtains the machine on January 2nd and the machine disintegrates into nothing on January 1st four years later, leaving nothing of value but also no special cleanup expenses. The machine costs one million and is not subject to technological or fashion changes, so MSO knows that every four years it will spend a million on replacing the machine.

The remaining income statement of MSO has revenue of $100 million and expenses of $90 million every year with no changes. Without depreciation, on the years that MSO has to buy the machine, MSO would have an additional $1 million of expenses. As a result, MSO's earnings would be $9 million every fourth year and $10 million the three intervening years, during which MSO would not have the machine-buying expenditure.

That the earnings of MSO would fluctuate despite that the environment is utterly constant and predictable, makes no sense. Neither the $10 million nor the $9 million are accurate figures of the true earnings of MSO. Obviously, MSO's true annual earnings are the average of the four years. The infrequent and large expense of the machine creates a predictable and correctable error.

The accounting solution is depreciation. Instead of having the income statement show an expense of $1 million every four years, the cost of the machine is spread over its useful life. The machine appears on the balance sheet as an asset and loses a quarter of its value each year. That loss of $250,000 appears as an expenditure on the income statement. The result is that the income statement becomes accurate and steady, showing earnings of $9.75 million each year, regardless whether it is a year that MSO buys the machine or not.

When we leave this stylized ideal of a machine with a perfectly predictable life, no residual value, and unchanging price and usefulness, depreciation is not as perfectly smooth a concept. For example, vehicles lose disproportionately more value during the first year than during subsequent years. Well-maintained equipment may survive well past the point when they are fully depreciated, i.e., their value will be depreciated down to zero but the equipment will still be providing use to the business. Replacements in fast-changing technologies may be more productive and less costly, with computer and telecommunication technology offering striking examples, where the replacements cost less and do much more than the original equipment. Each of those phenomena produces some slippage between the ideal of depreciation and its actual implementation. In each case, the accounting statements deviate from the ideal of showing the business's expenses accurately. Exercises 5–8 ask you to identify the biases that each one of these phenomena produces.

## 5    The Visual Balance Sheet

The visual approach to the balance sheet converts assets and liabilities to pieces of cloth hanging from two different clothespins on a single clothesline. On the left-hand side is the clothespin labeled *Assets* and on the right-hand side is the clothespin labeled *L&E* for *Liabilities and Equity*. Equity is a different type of cloth, an infinitely stretchy one, an extreme version of spandex, perhaps. To stress the stretchy quality of the fabric that corresponds to equity, it is yellow in the graphs. Going back to the MSO spreadsheet, the cloths hanging from the clothespin labeled *Assets* reach down by $309 million. The cloths hanging from

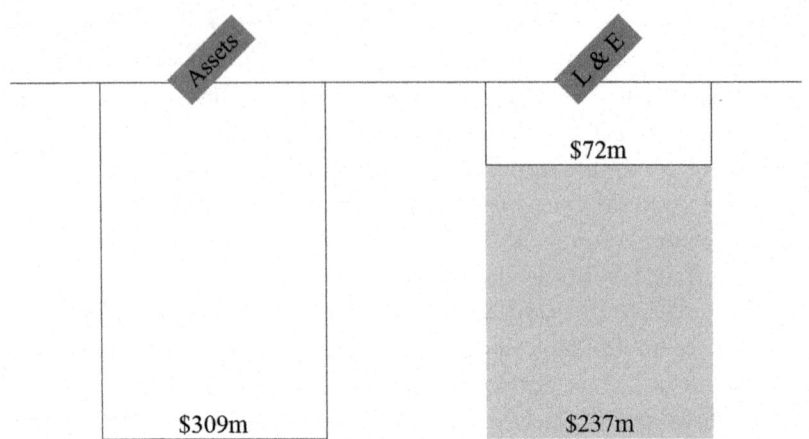

**Fig. 1** The visual balance sheet of MSO

the right-side clothespin reach down $72 million, corresponding to liabilities. Those are fixed and do not stretch or shrink. From there and down, the stretchy fabric gets connected and stretches down to reach the length of the assets. Therefore, it stretches to $237 million. Figure 1 illustrates.

The advantage of the visual balance sheet is that it reveals the relative size of the amounts at issue in the balance sheet. For example, consider the amount of current assets and liabilities and, as the next subsection will do, par value.

Current assets are $234.4 million and current liabilities $61 million. Figure 2 displays the result of separating out those two amounts from total assets and from total liabilities.

To the extent that accountants use the current amounts of the balance sheet to show viability, MSO appears unambiguously viable. The current assets are much larger than the current liabilities. MSO has virtually no risk of becoming unable to pay its obligations in the immediate future.

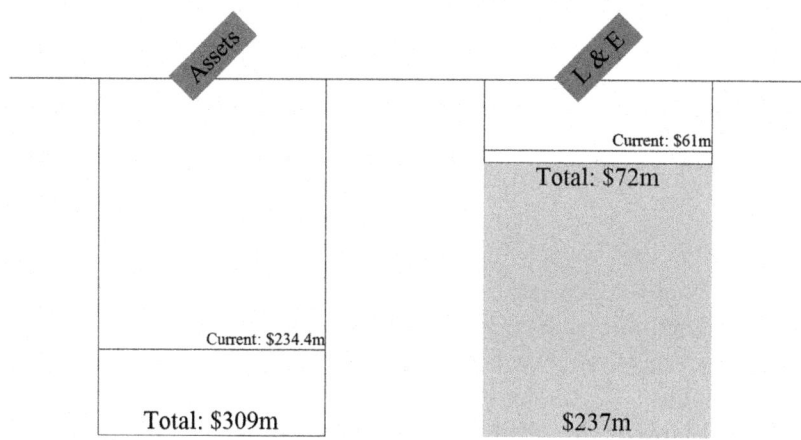

**Fig. 2**  The visual balance sheet with current amounts

# 6    A Seniority Balance Sheet

Because MSO is so financially healthy, it offers a poor example of how to form a seniority-based balance sheet for the distribution of its value to its claimants. To see such an example that includes a secured lender, assume that all the noncurrent assets of MSO secure an obligation to Bank for $250 million.

Observe, first, that MSO is insolvent in the sense that it has fewer assets than liabilities.[5] Second, observe that the bank's claim is undersecured, meaning that the bank's collateral, valued at $74.6 million according to the balance sheet (from total assets of $309 million minus current assets of $234.4 million) fails to cover the entirety of the bank's $250 million claim. Because the bank is undersecured, if the bank were to exercise its right to foreclose against the collateral, the sale of the collateral would only raise about that amount. The bank's secured status gives the bank priority against the proceeds of the sale of its collateral,

---

[5]This type of insolvency is also called *bankruptcy insolvency* or *balance-sheet insolvency*. These terms distinguish it from inability to service debts, which is also called *equity insolvency* or *cash-flow insolvency*.

**Table 2**  Seniority balance sheet of MSO

(all amounts in millions)

| ASSETS | | LIABILITIES AND EQUITY | |
|---|---|---|---|
| COLLATERAL | | SECURED CLAIMS | |
| Bank's Collateral | $74.6 | Bank's sec'd portion | $74.6 |
| UNECUMBERED ASSETS | | UNSECURED CLAIMS | |
| Remaining assets | $234.4 | Bank's deficiency claim | $175.4 |
| | | Other claims | $72 |
| | | Tot. Unsec'd. debt | $247.4 |
| | | EQUITY | none |

but that would still leave the bank unsatisfied by about $175.4 million ($250 million claim minus $74.6 million proceeds from the sale). While it is conceivable to have a contractual arrangement by which a secured creditor waives its right to be satisfied by other assets of the debtor (such loans are called *non-recourse* loans), that is unlikely and, for the sake of this example, it is not the case. The bank's *deficiency claim* for $175.4 million, then, would become an unsecured claim.

In the right-hand side of the seniority balance sheet, Table 2, claims are listed by, and grouped by, seniority. The classes of seniority are secured claims, unsecured or general claims, subordinated claims, and equity. Equity may further divide into preferred and common, with preferred being senior to common.[6] On the left-hand side, the assets that are collateral will appear on top. The result is a balance sheet with $74.6 million in collateral corresponding to the $74.6 million claim of the bank, and remaining assets of $234.4 million, corresponding to unsecured claims of $247.4 million, the $72 million of obligations of the real MSO plus the hypothetical deficiency claim of Bank for $175.4 million. In other words, the unsecured claims are greater than the unencumbered assets of MSO. If MSO were to be liquidated, then the unsecured creditors would all receive the same proportional satisfaction of 234.4/247.4ths of their claims or about 94.75 cents for each dollar of their claims.

---

[6]Do not confuse preferred and common stock with the two classes of stock that MSO has outstanding, class A and class B. It appears that both classes of MSO stock have the same seniority, whereas if one were senior to the other, meaning that in liquidation it would need to be paid in full before any value would go to the other class, then the senior class would correspond to preferred stock. Presumably, the reason that MSO has two classes is to give one class greater voting power, for the founder to maintain control of the board of directors with less than a majority of the stock.

The main assumptions of this example are that the assets of MSO would be valued at the same amounts at which the assets are listed on the balance sheet, and that MSO, as a business, is not viable. Because the MSO business is likely viable, the business could be sold for more than the liquidation of its assets. Also, for the same reason, MSO is a good candidate for a reorganization under Chapter 11 of the Bankruptcy Code, where further bargaining may occur. Nevertheless, the seniority balance sheet would form the baseline for the negotiations.

Remove the hypothetical $250 million obligation to Bank and return to the real MSO balance sheet. The next challenges are to understand par value and to see some basic transactions and how they would alter the balance sheet.

# 7    Par Value and Some Transactions

Notice that MSO has two classes of common stock. Class A has about 20 million shares outstanding with a par value of $0.01; class B has about thirty million shares outstanding with a par value of $0.01. This produces the par account and adding it to the visual balance sheet reveals its importance. Corporate law prohibits distributions and repurchases that would violate the par account. The par account is the result of multiplying one cent (the par value per share) by the number of shares outstanding, 50 million, which produces a par account of $500,000. At the image size that we are using, the par account is not visible. Figure 3 multiples the par account by 10 to make par value $0.10 per share and the par account five million and (barely) visible.

Note that, because the par account is part of the equity, it is yellow. Yet, it does not have the same stretchy nature as the overall equity does. If the corporation gains or loses an asset, the equity stretches to adjust, but the par account does not change. Vice versa, if additional stock is issued, redeemed, or bought back, the par account will change but the equity may not. Whether the equity changes depends on the fundamental equation of accounting, whereas the size of the par account depends on the number of shares outstanding.

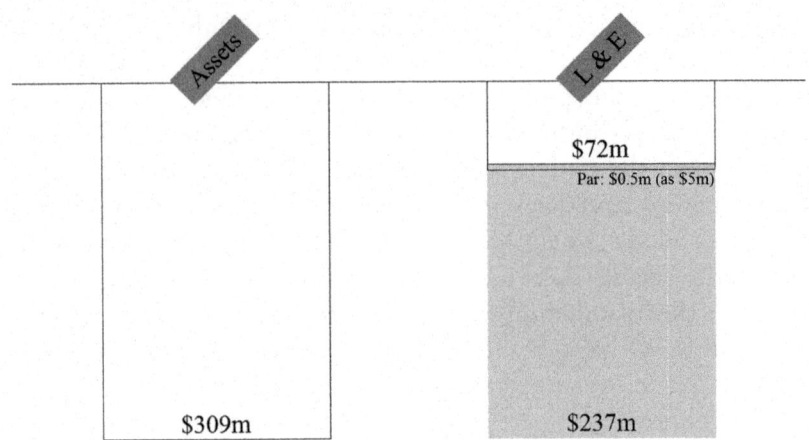

$72m

Par: $0.5m (as $5m)

$309m

$237m

**Fig. 3** MSO with a par account inflated ten times

With this balance sheet and the par account at $5 million, MSO can distribute a dividend of up to $232 million; dividends cannot be distributed out of the par account. This tiny par account reflects some features of modern corporate practice. (1) The creditors do not rely on the par account as an important protection but negotiate their own protections. (2) Investors in stocks do not feel that a high par value is desirable. (3) The ease of amending corporate charters under modern corporate law makes the protection of the par account weak and subject to easy amendment.

A historical comparison is revealing. Moody's 1921 entry (a) on Ingersoll-Rand Corporation shows a balance sheet with almost eleven million par account for the common stock where total liabilities and equity are under $41 million; (b) on Carnation Milk Products Corp. shows a balance sheet with nine million par account where total liabilities and equity are $17.2 million.[7] In such balance sheets, the par accounts truly offer significant protection to creditors. Yet, leafing through the 1921 Moody's we already see many of the best-known corporations, such as Ford and GM, having reincorporated in Delaware and having issued no-par common stock.

---

[7] 1 Moody's Manual of Railroad and Corporation Securities 1632–33, 1764–65 (1921).

Let us observe how the balance sheet would change if MSO undertook two alternative transactions. In the first alternative, MSO borrows $20 million and distributes the proceeds as a dividend. In the second alternative, MSO issues five million additional shares of $0.10 par value stock at the current market price on March 4, 2004, of $10 per share.

Consider first the borrowing transaction. The instance after MSO has received the loan, MSO will have $20 million of additional liabilities and $20 million of additional assets. This makes total liabilities $92 million but does not change current liabilities, assuming this is not a very short-term loan. Then, MSO distributes a dividend of $20 million in aggregate, removing the new assets from its balance sheet.

The equity changes because of the amount of dividend distributed. The equity is the result of applying the fundamental equation of accounting: assets of $309 million minus liabilities of $92 million give equity of $217 million.

Figure 4 is the result. The current assets and liabilities appear again. The end result is that debt increased by $20 million and equity shrank by that amount with no change on the asset side—there, the money came and went, temporarily increasing current assets and total assets, but that increase disappeared when the funds were dividended out.

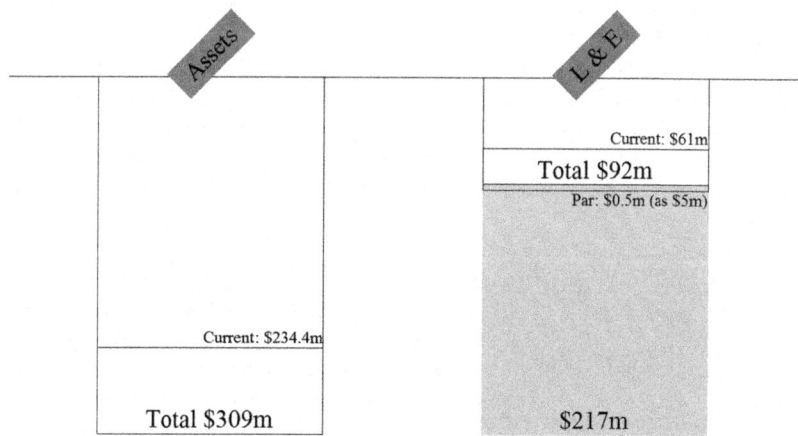

**Fig. 4**  MSO after taking a $20 million loan and dividending it

The next transaction has MSO issue five million additional shares, for the current market price of $10 in cash. This produces two adjustments. We need to establish the amount of cash that MSO will raise, and we need to adjust the par account for the additional shares. MSO will raise cash of $50 million by this issuance. Since this is in exchange for cash, the new asset is a current asset and will also increase current assets to $284.4 million, as well as total assets to $359 million. Keep in mind that if the new stock had been issued in exchange for a long-lived asset, then the increase would not accrue to the current assets.

The number of shares outstanding will increase from 50 million to 55 million. Using our hypothetical $0.10 par value per share (rather than the actual one cent), the par account will increase from $5 million to $5.5 million. The equity will stretch to accommodate the new amount of assets. Using the fundamental equation, the new assets of $359 million minus the liabilities of $92 million give the new equity of $267 million. Figure 5 is the result.

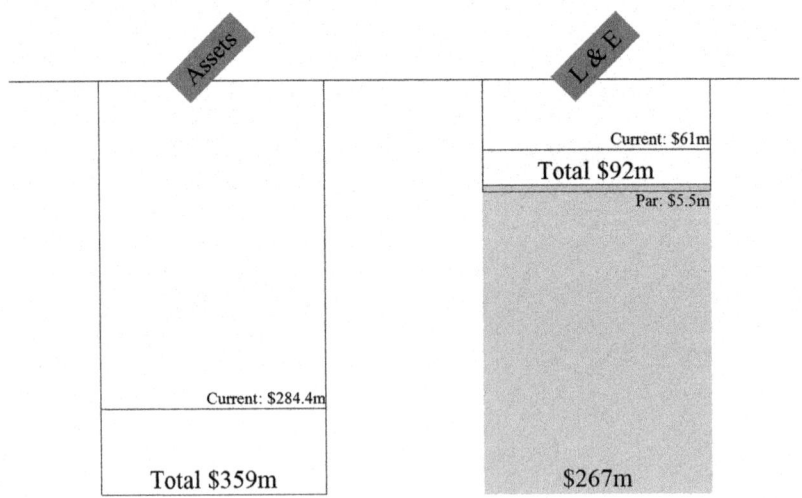

**Fig. 5**  MSO after the loan and a stock issuance for cash

Taking into account the current market price of MSO stock, however, reveals that this balance sheet does not correspond to how the stock market values MSO.

# 8    The Enterprise-Value Balance Sheet

Go back to the actual balance sheet of MSO and consider that it has 50 million shares outstanding in two classes: one with about 20 million shares outstanding, and one with about 30 million shares outstanding. Suppose the two classes of stock have the same value (in reality one class has probably greater voting rights so it would be more valuable, but we ignore this for now). Those shares have a price of $10 on the day that this balance sheet is issued (and the balance sheet is not outdated[8]). This means that the market believes that the equity of MSO has a value of about $500 million (this is called the *market capitalization* of MSO) while the balance sheet indicates that the equity of MSO is $237 million or, per share, $4.74 (this is called the *book value* of MSO, per its accounting books). The market values MSO more than twice its book value. How can that be?

Recall the Capital Asset Pricing Model (Chapter 5). How are investments, such as MSO stock valued? The CAPM tells us that investments are valued by taking into account their beta (their measure of system-wide risk) to establish the proper expected return and discounting the forecasted expected return using that rate. Although investors may not be able to do that as scientifically as the CAPM suggests, the CAPM shows the primary drivers of valuation are risk and expected return. The actual assets of the business are irrelevant! Internet companies may be the extreme examples of this discrepancy between market capitalization and book value. An internet

---

[8]The balance sheet is dated December 31st but is issued on March 4th. For the market price to match the balance sheet, the MSO balance sheet would have to experience an increase in its assets of about $263 million, a virtual doubling of MSO's assets, from December 31st to March 4th, which has not happened.

company like Google or Facebook has as primary assets sets of server banks; those have small value compared to the value of the enterprise, which is driven by the profit that investors expect Google or Facebook to make.[9] Similarly, for MSO, the profit that investors expect MSO to make (and MSO's risk) lead them to be willing to buy it at $10 per share and lead to an implied valuation of its equity at about $500 million rather than the $237 million of equity that its balance sheet shows.

The primary lesson is that valuation follows projected gains, not the past that the financial statements reflect. However, because the balance sheet offers useful intuitions, the balance sheet is worth adjusting for the enterprise valuation, when that is known, as in publicly traded stocks, or for an estimation of the enterprise valuation, as has to be done in other businesses. Essentially, the question becomes: Given the value of the claims against the company, what is the value of the enterprise? This is a reversal of the fundamental equation of accounting. Instead of trying to derive the equity from taking the assets as given and subtracting the face amount of the liabilities, we take the debt and the market capitalization of the stock (and any other classes of securities, such as bonds and preferred stock) and we derive the implied value of the enterprise.

In the case of MSO, deriving its enterprise value is straightforward. The liabilities side of the enterprise-value balance sheet will have the current liabilities of the accounting balance sheet ($72 million) and the equity at its market capitalization ($500 million) for a total of $572 million. The resulting asset side of the enterprise-value balance sheet will have enterprise value of $572 million.

---

[9]For example, Facebook on July 26, 2017, had a market capitalization of $479 billion, about eight times its equity of $59 billion (according to the *Wall Street Journal*, overview: perma.cc/ER3V-GZTH; balance sheet: perma.cc/R7NQ-FWSE); similarly, Google (Alphabet) had market capitalization of $664 billion, about four-and-a-half times its equity of $140 billion (overview: perma.cc/X5FJ-X6YP; balance sheet: perma.cc/6AG3-MGGM).

# 9    A Fraudulent Transfer Law Application

The choice of whether to apply the enterprise-value balance sheet or a balance sheet based on the value of the assets (not quite the accountants' balance sheet, as that might have additional values that do not correspond to real assets as we will see in merger accounting) can be crucial in the application of fraudulent transfer law in leveraged buyouts. A leveraged buyout is the acquisition of a business where the acquirer uses a lot of debt (leverage) to finance the acquisition. The new debt may give rise to fraudulent transfer claims on behalf of existing creditors. Fraudulent transfer law turns on the solvency of the debtor business at the time of the buyout. Often, applying the enterprise-value balance sheet will indicate the business is solvent, because its expected profits produce an estimated enterprise value greater than the liabilities, whereas summing the value of the assets does not.

Courts determine which of the two to apply based on whether current assets exceed current liabilities. If current assets exceed current liabilities, then the business is viable and the enterprise-value balance sheet is appropriate. If the current assets are short of the current liabilities, then the viability of the business is doubtful; therefore, the assumption that future profits will materialize is dubious and the discounting of future profits makes little sense. Then, courts tend to use the asset-value balance sheet, which tends to indicate insolvency, because the values of the assets tend to sum to less than the debt burden.[10]

# 10    Merger Accounting

Accounting has a particular problem in mergers and acquisitions. A business that buys a second business will necessarily pay a premium over the market valuation of the stock, usually a premium between 15 and

---

[10]*See, e.g.*, *In re* O'Day Corp., 126 B.R. 370, 403 (Bankr. D. Mass., 1991) ("[I]f reasonable projections would have shown losses in the years after the LBO, ... a balance sheet approach is appropriate [as opposed to valuing the going concern]" with very detailed analysis and numerous further citations).

50% compared to market prices before the announcement of the acquisition. If the acquirer would not pay a premium, then the shareholders of the target would have little reason to sell. Thus, the acquirer may be forced to share the gains that the acquirer expects to make from the target, the gains, for example, from running it better, or from the synergies that the acquirer expects to produce with the acquirer's existing businesses, etc. When the accountants turn to create a single balance sheet for the combined businesses, the fact that the acquirer has bought the target at a market transaction makes the accountants need to use that valuation. However, that is in tension with the typically much lower asset prices that the target's balance sheet carries. The solution involves, first, marking to market both sides of the target's balance sheet, and, second, the creation of a fictitious asset, called goodwill, that makes the balance sheet balance again. This is called the *purchase* method and is the only one that accounting rules allow—they used to also allow merging entities to use the *pooling* method, which ignored the valuation of the target in the transaction and maintained the existing entries on the balance sheets.

The purchase method of accounting for mergers does not change the asset values on the balance sheet of the acquirer. However, the right-hand side of the acquirer's balance sheet may change as the terms of the transaction require, i.e., if additional debt or equity is issued. Moreover, when equity is issued in the context of the transaction, that equity's market value must appear in the equity of the acquirer. For example, if the acquirer issues shares to the target with a value of $1 million, that million will appear as additional equity on the equity of the acquirer, perhaps broken up between the increase of the par account and a different equity entry.[11]

---

[11]This stands in contrast to outstanding shares of the acquirer. Those likely already trade at a price significantly higher than their book value, but that does not trigger any obligation to update the balance sheet. In a sense, this treatment of the newly issued shares is not a result of the treatment of the acquirer's balance sheet but of the marking to market of the target's balance sheet. The target's equity gets revalued at the value of the consideration that the acquirer gives. Then, because the consideration is equity, it appears as such on the acquirer's balance sheet.

The target's balance sheet, however, is rebuilt. Every asset and every liability is marked to market, taking its current market value. Then the equity is replaced according to the terms of the transaction and takes the value implied by the transaction. The result is that the right-hand side of the balance sheet is likely larger than the left-hand side, i.e., the marked-to-market liabilities and the equity priced according to the deal value will tend to be greater than the value of the assets (despite that the assets are also marked to market). This produces a balance sheet that does not balance. To make the balance sheet balance again, accounting rules require the creation of an additional, fictitious asset. The asset is called goodwill, but it does not correspond to any actual intangibles that the acquired business may have. To the extent accounting rules recognize any such intangibles and a market value for those can be estimated, those would have appeared separately as assets in the marking-to-market process. Rather, this goodwill is nothing but an accounting entry designed to make the balance sheet balance again. Thus, instead of taking the assets and liabilities as fixed and using the fundamental equation of accounting to derive the equity, the purchase method takes liabilities (marked to market), equity (valued according to the deal), and assets (marked to market) and uses goodwill to balance the balance sheet.

To observe an example of the operation of the purchase method of accounting for mergers, let us construct a target with outstanding debt that is publicly traded. The debt trades at a price lower than its face value (perhaps interest rates are higher than at the time of issuance, or the risk of the debtor has increased). Let us use as the target a version of MSO that has issued $80 million of bonds, which trade at a 25% discount from face value, i.e., the bonds trade at an aggregate of $60 million. To maintain the same total enterprise value as the real MSO has, we subtract the hypothetical debt from MSO's actual market capitalization of its stock to obtain a hypothetical market capitalization of $440 million (the result of subtracting $60 million of hypothetical bonds at their market value from MSO's actual capitalization of $500 million). Accordingly, each of the 50 million shares of MSO trades at $8.80 before the announcement of the merger according to this hypothetical (the result of dividing the market capitalization of $440 million by the 50 million shares outstanding).

Let us juxtapose the enterprise-value balance sheet to the asset-value balance sheet. The left-hand side of the asset-value balance sheet will be identical to that of the actual MSO seen in Table 1. The asset-value balance sheet will have $234.4 million in current assets and total assets of $309 million, implying noncurrent assets of $74.6 million. The right-hand side of the balance sheet will adjust for the additional debt. Thus, below the current liabilities of $61 million, will appear the $80 million of bonds at face value, making total liabilities $141 million. Applying the fundamental equation of accounting means that the equity will be $168 million (assets of $309 million minus liabilities of $141 million). This is circumventing the question on what exactly the intangible assets of $44.2 million are on the balance sheet of Table 1. If those are the value of patents or other true intangible property, then they correctly should appear in the marked-to-market balance sheet. Proceed on the assumption that those intangibles are true assets, not accounting goodwill from prior acquisitions by MSO, and marking to market will not change them.

The acquirer, we will assume, will pay $10.56 per share, a 20% premium over the $8.80 market price for MSO.[12] We will assume that the acquirer has total assets of $720 million with current assets of $25 million leaving $695 million of noncurrent assets, current debt of $87.5 million, long-term debt of $295.5 million, making its equity $337 million; that the acquirer has 40 million shares outstanding with a par value of $1 per share and, therefore, a par account of $40 million. (Some rounding occurs throughout these calculations.)

---

[12]It turns out that a couple of years later, MSO was bought out at a 20% premium over its market price at that time. However, in the meanwhile Martha Stewart was convicted and spent some time in jail and MSO itself had not done well so that its market price was quite a bit lower. *See,* Michael J. de la Merced, *Martha Stewart's Empire Sold for Fraction of Its Former Value,* N.Y. TIMES June 22, 2015, at B6 https://www.nytimes.com/2015/06/23/business/dealbook/martha-stewarts-media-empire-sold-for-fraction-of-its-former-value.html [perma.cc/5R5A-CWN5] (announcing MSO being bought by Sequential Brands for $6.15 per share for a total price of $353 million suggesting the outstanding shares were about 57.4 million).

The terms of the transaction are that the shareholders of MSO will receive half the consideration in shares of the acquirer and half in cash; that the exact number of shares to be received will be determined by the average market price of the acquirer's shares at the close of trading the last five trading days before the actual merger. Also, the acquirer has obtained a loan commitment from a lender for half the purchase price—this loan is the source of the cash that the acquirer will pay for the 50% of the purchase price.[13] Name the acquirer Sequential Brands. The average closing price of Sequential Brands the last five days before the merger turns out to be $7.00.[14] As a result, the number of Sequential Brands shares issued for each MSO share are about .754 (to produce a consideration of $5.28 per MSO share, half the purchase price of $10.56 divided by the price of Sequential Brands shares of $7— the other half of the consideration is cash). Therefore, Sequential Brands will issue 37.7 million shares of Sequential Brands in aggregate for the 50 million of MSO shares outstanding. Added to the already outstanding 40 million shares of Sequential Brands, the result is the combined entity will have a total number of shares outstanding of 77.7 million. The new par account of the combined entity will be $77.7 million. The par account increases by $37.7 million due to the additional shares being issued to the MSO shareholders. However, the market value of those shares is $7 each. Thus, additional value (as equity) is being paid beyond the increase of the par account, additional value of $6 per issued share, in total $226.3 million ($6 times the 37.7 million shares being issued). The consideration that the equityholders of MSO receive in

---

[13]These terms have some similarities to the deal between MSO and Sequential Brands, which was a 50% cash deal based on the average price of Sequential on the last few days before the merger but gave each shareholder the option to elect away from the 50–50 split subject to proration; was for a 20% premium over the pre-announcement price (but the resulting price was $6.15 per share); was to be paid with loan proceeds; at a time when Sequential had about $295.5 million of long-term debt and about 30 million shares outstanding but they had par value of $0.01 per share but traded in the $12–$15 range. *See generally* Martha Stewart Living Omnimedia, Inc., Proxy Statement (Schedule 14A) (Oct. 27, 2015) www.sec.gov/Archives/edgar/data/1091801/000114420415060980/v422845_defm14a.htm [perma.cc/68ES-H5F7].

[14]This figure is approximately the price of Sequential Brands stock in March of 2013.

equity of the merged company is thus $264 million, $37.7 million in equity corresponding to par and $226.3 million as additional equity.

The additional debt that the combined entity will incur is the source of the cash for the other half of the consideration for the MSO shares. At $5.28 per MSO share multiplied by the 50 million MSO shares outstanding, the additional debt will be $264 million.

The right-hand side of the balance sheet after the purchase of MSO will, therefore, have (1) as current liabilities (a) those of MSO (marked to market) at $61 million and (b) those of Sequential Brands (at book value) at $87.5 million; plus, (2) as noncurrent liabilities, (a) those of MSO (marked to market) at $60 million, (b) those of Sequential Brands (at book value) at $295.5 million, (c) plus the new takeover debt of $264 million. (3) Equity follows those with (a) the par account of $77.7 million, (b) additional equity value received by MSO shareholders of $226.3 million, and (c) the original additional equity of Sequential Brands (at book value) at $297 million (old equity value of $337 million minus the equity corresponding to the par account of $40 million). This gives a total liabilities and equity side of $1369 million.

New goodwill will be created to make the assets reach this amount. On the assets we have (1) current assets consisting of (a) MSO current assets (marked to market but unchanged) of $234.5 million, and (b) Sequential Brands current assets (at book value) of $25 million; (2) noncurrent assets of (a) MSO (marked to market but unchanged) at $74.5 million, and (b) Sequential Brands (at book value) at $695 million. This gives total assets before goodwill of $1029 million. Subtracting this from the total liabilities and equity means that the new goodwill must be $340 million.[15] After adding this goodwill, the balance sheet balances again. Table 3 has the resulting balance sheet of the combined entity.

The purchase method does not change the acquirer's assets and liabilities; it leaves them at the traditional accounting treatment (a) of historical cost for the assets (with any depreciation adjustments) and (b) of face

---

[15]This is new goodwill as opposed to potentially existing old goodwill due to prior acquisitions of Sequential Brands. The assumption, however, is that no old goodwill exists.

**Table 3**  Balance sheet of combined entity

```
        (as of March 5, 2004, in million, except per share amounts)
ASSETS                                  LIABILITIES AND EQUITY
  CURRENT ASSETS                          CURRENT LIABILITIES
    MSO current assets      $234.5          MSO current liab.          $61
    SeqBr current assets     $25            SeqBr current liab.        $87.5
      Total Current Assets  $259.5            Total Current Liab.     $148.5
  NONCURRENT ASSETS                         NONCURRENT LIABILITIES
    MSO noncurrent           $74.5          MSO bonds                  $60
    SeqBr noncurrent        $695            SeqBr debt                $295.5
    Goodwill (new)          $340            Takeover debt             $264
      Tot. Noncurr. A.    $1,109.5            Tot. Noncurr. debt      $619.5
        Total assets      $1,369       EQUITY
                                         Par account (77.7 mill. shares of
                                           par value $1 per share) $77.7
                                         Further equity value associated with
                                           the MSO acquisition    $226.3
                                         Original further equity of SeqBr
                                                                    $297
                                             Total Equity           $601
                                             Total L&E             $1,369
```

value for the liabilities. The radical change happens to the balance sheet of the target, here MSO, in which assets and liabilities are marked to market and the equity of which gets replaced by the consideration, debt or equity that the acquirer issues in the acquisition, with corresponding consequences for the resulting par account of the acquirer. The marking to market of the assets and the creation of goodwill (which accounting rules used to require to be amortized down to zero over a large number of years but now accounting rules leave goodwill unamortized) tend to increase depreciation and amortization expenses. At earlier times, when accounting rules gave a choice of using the purchase method or a different method for accounting in mergers, many firms preferred not to use the purchase method in order to present to investors greater earnings (due to avoiding the larger charges for depreciation and amortization of the purchase method).

Marking liabilities down to their market value appears counterintuitive but it makes sense. If the business can buy a liability back for less than its face value then that is the rational course of action, not paying the face amount. In this example, the combined entity owes $80 million on the MSO bonds in terms of face value. It seems wrong to bring that debt on the combined balance sheet at $60 million just because it trades at that value. However, the business can borrow at the market rate and buy back the bonds. The business would only need to borrow

$60 million to buy back the bonds, despite that their face value is $80 million. Therefore, the true indebtedness is for the market value of the debt, $60 million, rather than the $80 million face amount.

Perhaps using visual balance sheets clarifies the accounting treatment of mergers. The acquirer's balance sheet remains in its asset-based mode. However, the target switches from an asset-based balance sheet to one based on enterprise value. Then the acquirer adds to the right-hand side of the acquirer's (asset-based) balance sheet the right-hand side of the target's (enterprise-value) balance sheet. The target's enterprise value comes from the valuation according to the acquisition, i.e., the market value of the consideration that the target's shareholders receive plus the market value of other claims. On the asset side of the acquirer's balance sheet, the purchase method adds the assets of the target and the remaining enterprise value of the target is added as goodwill.

Figure 6 shows the visual balance sheet of MSO using enterprise value, using as value of the equity what the equityholders will receive according to this transaction, $528 million. The debt of MSO appears

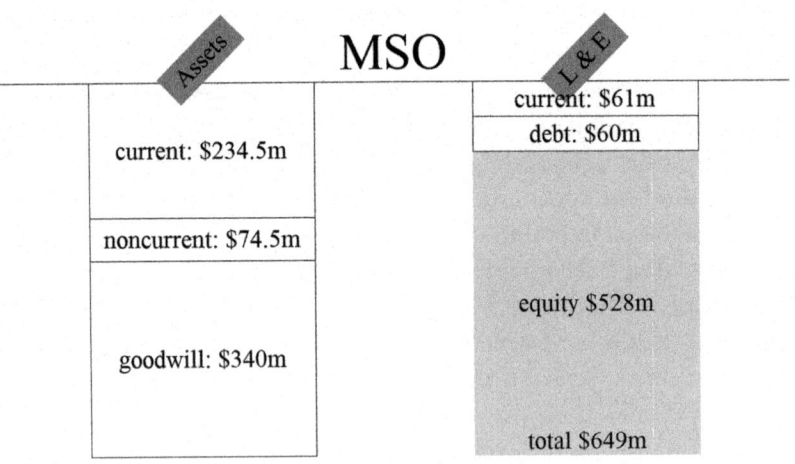

**Fig. 6** MSO enterprise value according to the terms of the deal

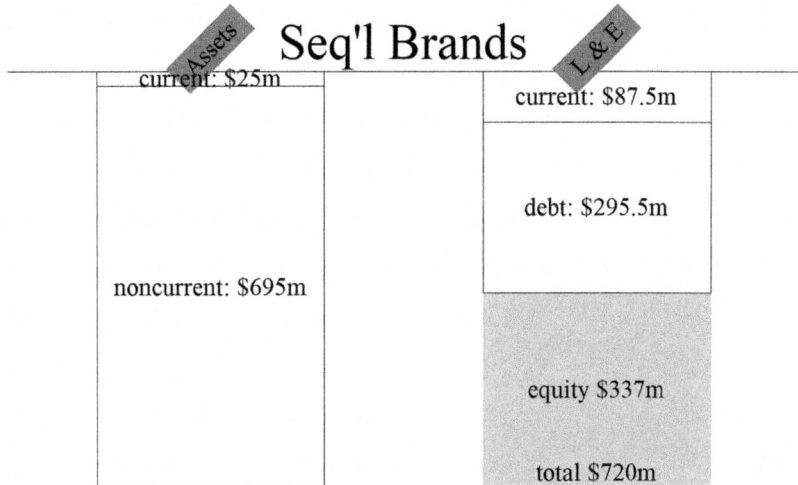

Fig. 7 Sequential Brands before the acquisition

at its market value of $60 million. For the enterprise-value balance sheet to balance, we need to add goodwill on the asset side. That has to be $340 million. Keep in mind that the total of either side of this enterprise-value balance sheet is $649 million.

Figure 7 shows the balance sheet of Sequential Brands before the acquisition of MSO. This is not based on enterprise value. Rather, the balance sheet of Sequential Brands is based on asset values. Notice that the total of either side of this balance sheet is $720 million.

Figure 8 shows the result of the combination of the two balance sheets into the post-acquisition balance sheet of the combined entity. This balance sheet is the result of summing the corresponding entries in the last two balance sheets. The total, $1369 million, is the result of adding the total of the MSO balance sheet, $649 million, to the total of the Sequential Brands balance sheet, $720 million. However, the debt and the equity are not the result of adding because additional debt arose from the acquisition, for the half of MSO's equity exchanged for cash. Thus, the debt is the result of the debt of each company plus the debt

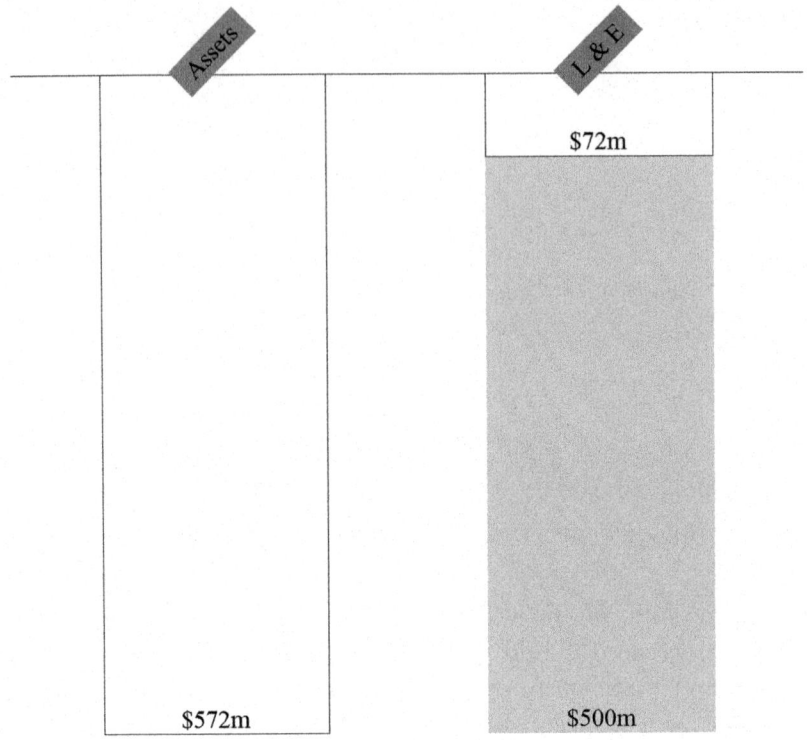

**Fig. 8**   The merged entity's balance sheet

issued for the acquisition. The new equity is the result of adding the other half of MSO's equity to the equity of Sequential brands. The new debt is the sum of the debt of the two businesses plus the additional debt issued as part of the deal.

The visual balance sheets allow a different perspective on the purchase method of accounting for acquisitions. In this example, the right-hand side of the acquirer's balance sheet grows by the enterprise value of the target. The target's enterprise value results from the valuation of the target according to the terms of the deal. This use of enterprise value creates a gap on the asset side of the balance sheet. Goodwill bridges that gap.

# 11   Conclusion

Financial statements are the language of business. Financial statements are an elaborate technological tool to describe the affairs of businesses. A basic understanding of financial statements is necessary for interacting with businesses. This chapter tried to help this basic understanding and proceeded to one more complex application, the handling of acquisitions on the balance sheet.

# 12   Exercises

1. The first few exercises ask you to track the balance sheet and income statement (and the visual balance sheet) of a simple corporation through some transactions. You can think of the corporation as a cookie jar; items enter the jar or exit from it, and the balance sheet and the income statement display the results. The first transaction is the funding of the corporation. The owners of the stock, who agreed to form the corporation, pay $1000 per share to acquire the corporations initial hundred shares. What is the balance sheet?
2. The corporation acquires a truck by paying $5000 and borrowing the remaining $15,000 from the manufacturer. What is the balance sheet?
3. The corporation uses the truck to provide moving services to four homeowners. The corporation hires two individuals as movers for 10 hours each at $15 per hour and charges the homeowners $400 each. The corporation sent out the bills on Monday of last week and has received payment from two homeowners. What is the balance sheet?
4. The remaining two homeowners pay their bills. The first quarter of the corporation's existence comes to an end. The rent for the corporation's premises for this quarter was $100. What is the income statement for this first quarter (assuming that the truck is depreciated over 10 years on a straight line and the truck's final value is zero)? What is the balance sheet?

5. The value of the truck after three months of use is $11,000. How is that different from its value according to the corporation's depreciation scheme?

6. Ten years go by, and due to good maintenance of the truck, it is still operating well, and is expected to be operational for at least five more years. How are these circumstances in tension with the depreciation scheme of the corporation? What does this mean about the true operating costs of the corporation?

7. Suppose that during the 1950s, the machines that car manufacturers used to make cars lasted exactly 5 years and would be replaced with a machine with the same productivity and at the same cost. Suppose accordingly, that these machines cost $100,000 and the remaining items in the income statement of a car manufacturer produced a profit of $80,000. The car manufacturer needs to buy one machine every five years. What would the manufacturer's income be each year of a six-year cycle without depreciation, and what would it be with depreciation?

8. Suppose that during the 1990s, office computers lasted three years and the computers with which they would be replaced would cost 30% less and be 30% more productive than the computer they would be replacing. How does depreciation work differently than in the prior exercise?

9. Suppose the terms of the MSO acquisition by Sequential Brands were identical in every way to the description in the text except for the part that Sequential Brands would pay the half of the purchase price by borrowing. Instead of borrowing, Sequential Brands pays that amount to the MSO shareholders out of the cash (the current assets) of the combined entity. Create the balance sheet of the combined entity.

# 10

## Aversion to Risk

## 1 Risk Aversion

Attitudes toward risk are central to the choices people make. Therefore, attitudes toward risk are important for the incentive scheme that law and policy produce as well as for the assessment of values that finance studies. In this chapter, we enter the mathematical understanding of aversion to risk when it applies to money. Attitudes toward risk outside monetary choices are still not well understood. Also, attitudes toward risk about money do not seem closely related to those about unrelated activities. We see little paradoxical about a solo ice-climber, for example, who would pay for a mistake half-way up a frozen waterfall with the ice-climber's own life, in taking no financial risk and only investing in treasuries. Vice versa, a venture capitalist, most of whose investments are destined to fail, would not be seen as strange if the venture capitalist takes very few physical risks, driving exceedingly cautiously and exercising only in the safest ways. Aversion to physical risk has little relation to aversion to financial risk, and our economic models only deal with financial risk. From here, risk refers only to financial risk.

© The Author(s) 2018
N. L. Georgakopoulos, *Illustrating Finance Policy with* Mathematica,
Quantitative Perspectives on Behavioral Economics and Finance,
https://doi.org/10.1007/978-3-319-95372-4_10

The road to comprehending risk begins at simple gambles. Someone faces a choice between receiving a certain amount and two risky amounts. For example, the risky amounts may be $100 and $900, say, depending on the toss of a coin. If the coin comes heads, take $900, if tails then $100. The uncertainty, thus, is the $100 or $900 outcome. This example will be the foundation for understanding reactions to risk. The first step is to recognize the subjectivity of the correspondence between the risk and a certain amount.

## 2     The Certainty Equivalent

What is the certain amount at which one would surrender or sell the risk of the $100-or-$900 50-50 gamble? The rare risk-neutral person would not surrender the gamble for less than its average payout of $500. A risk-averse person would gladly give it up for less than that. As aversion to risk increases, smaller certain payments become acceptable substitutes for taking the gamble. The certain amount is called the *certainty equivalent* of the gamble.

Various parameters tend to influence the relation between the certainty equivalent and the expected value of the gamble. Naturally, the degree of aversion to risk and the size of the gamble are the most important and best understood from a quantitative perspective. For the same person, a smaller gamble will be much less undesirable compared to its expected value. The extreme example is a 50-50 gamble to get $499 or $501, the certainty equivalent of which cannot be below $499.

## 3     Models of Risk Aversion

Particular importance, however, takes the individual's wealth. Intuition and some evidence suggests that wealth influences attitudes toward risk. The 50-50 gamble of $100 or $900 has different importance for someone who is destitute compared to someone with great wealth. Winning could enable the destitute to afford things previously out of reach but would not change the life of a wealthy person. The idea that, regardless of wealth, people have the same attitude toward gambles seems counterintuitive; such a phenomenon would be *constant absolute risk aversion* because the aversion toward the risk would be constant as wealth changes as long as the risk remains the same in absolute terms. An appealing alternative is to consider that the attitude toward gambles remains constant as wealth changes if the gamble remains a constant ratio of the individual's wealth. This phenomenon is called *constant relative risk aversion* because aversion toward risk stays constant as wealth changes provided that the gamble stays constant relative to the changing wealth.

In mathematical terms, the description of the phenomenon implies that the certainty equivalent stays a constant fraction $t$ with respect to wealth $w$ as long as the gain and the loss stay constant fractions $f$ of wealth $w$. In algebraic terms, we could say that the certainty equivalent $tw$ is equal to the wealth that produces the expected utility of wealth, which in the case of a 50-50 gamble is the average of the two utilities, of ending up with wealth $(1-f)w$ or $(1+f)w$. This analysis forms the equation $u(tw) = [u((1-f)w) + u((1+f)w)]/2$. The left-hand side is the utility of wealth from having wealth equal to the certainty equivalent. The right-hand side is the average of the two utilities of wealth, that of losing a fraction $f$, and winning the same fraction. Because the risk is 50-50, instead of multiplying each outcome by its probability, their sum is divided by two.

Figure 1 illustrates the phenomenon of risk aversion as a consequence of the shape of the utility of wealth. The horizontal axis represents wealth and the vertical axis represents utility. The plotted function is a utility-of-wealth function that produces constant relative risk aversion. The figure displays the analysis of the 50-50 gamble producing wealth of either $(1-f)w$ or $(1+f)w$, i.e., losing or winning a fraction $f$ of wealth $w$. The crux lies in the comparison of the (greater) utility of having the expected wealth, $w$, to the (lesser) expected utility of the gamble, $u(tw)$. The wealth that produces the same utility as the gamble, $tw$, is the certainty equivalent; the full phrase would be that it is the certain wealth that produces utility equivalent to being exposed to this risk.

The numerical values that form the figure make a concrete example. The expected wealth of the individual who is subject to the risk is 3 (or

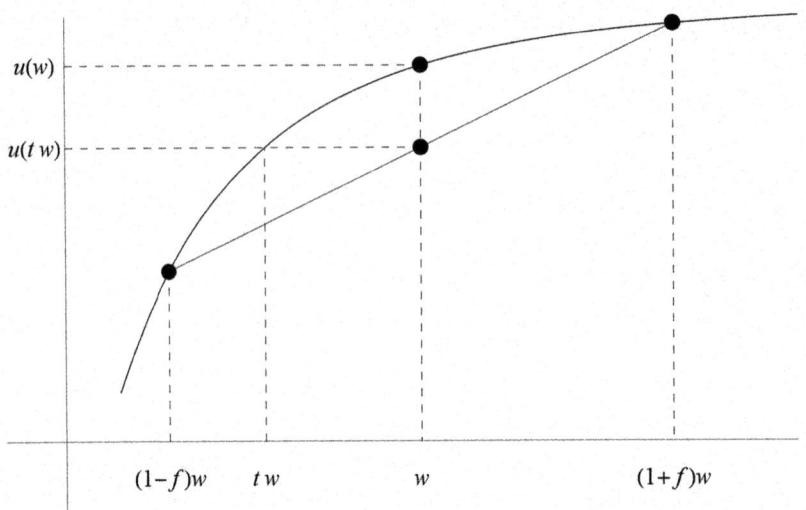

**Fig. 1**  Risk aversion as a result of the utility of wealth

$300,000 if we make the unit a hundred thousand dollars) and the fraction of the wealth at risk is $f = 1/3$, for a gain or loss of $100,000. The utility of wealth function that exhibits constant relative risk aversion with a coefficient of risk aversion of 4.5 becomes $u(w) = -.2857/w^{3.5}$. The certainty equivalent of the 50% probability of the utility of having $200,000, $u(2)$, and 50% probability of the utility of having $400,000, $u(4)$, corresponds to about $u(2.38)$ or having about $238,000, quite a bit less than the average wealth of $300,000. In other words, the happiness of having $400,000 with a 50% probability averaged against the 50% chance of having $200,000 makes this person feel as happy as if having $238,000, which corresponds to $u(tw)$ on the graph.

# 4    Insurance

Consider insurance from this perspective. Whereas the risk that an individual faces is strongly undesirable, an insurance company can aggregate the risks of many individuals and, instead of facing more risk, due to diversification, the insurer will face less risk proportionately. Effectively, each individual would see facing the risk as a loss of welfare and would gladly pay to have the insurer shoulder part of the risk. The insurer, due to diversification, considers the risk as less undesirable than the individual did, and is, therefore, willing to bear it at a price individuals are willing to pay.

To put it in the mathematical terms used immediately above, someone with wealth $w$ facing that uncertainty is willing to pay up to the reduction of wealth experienced due to the risk, namely $w - tw = w(1 - t)$ in order not to be exposed to this risk. The shape of the risk is a 50-50 chance to lose or double a sizable fraction of the individual's wealth. An insurer who can aggregate many of these risks so as to bear less risk (see the discussion on diversification, Fig. 2 in Chapter 5 and accompanying text, above) can charge each insured up to that amount to shoulder each risk. Suppose the insurer aggregates a hundred of these risks. What distribution does the risk that the insurer faces have? Whereas each individual faces a 50-50 double or nothing risk, the insurer, assuming that the risks are truly independent, faces a risk akin to having divided a purse among a hundred different cointosses.

The assumption of independence is strong, however. If we think of an insurer of homes against fire, the possibility of fire spreading from house to house in a region means that if the insurer covers several houses in a region, then their risk of fire is not independent. To preserve independence, the insurer must cover homes that are far apart, in different neighborhoods and cities. Fire is also much more rare than a 50% chance. A more appropriate model would be insurance on ships undergoing a particularly risky journey through, say, a region at war.

Suppose that the insurer manages to cover a hundred such risks. The insurer's experience follows that of the gambler who has divided a purse among a hundred different cointosses. As we saw, the corresponding distribution approaches the normal but is the binomial one for a hundred trials with .5 probability of success in each.

However, a problem arises from the possibility of insolvency of an individual insurer in the case that a disproportionately large fraction of the outcomes are undesirable. Therefore, let us posit a mutual insurance society, where each insured puts some amount as the capital plus the premium, expecting to receive back what is not spent making whole the insured. Each insured receives coverage and in most cases will receive back some capital. However, in the case that a large fraction of outcomes are bad, then the insurer becomes insolvent. In that case the capital is depleted. The insured who suffered the harm receive a prorated payment, because the insurer cannot pay all the damage. The remaining risk, beyond the prorated payment, is borne by the individuals.

The complete model, accordingly, must track, for each possible number of winning risks, from zero to one hundred, the expected utility of wealth of the insured who do not suffer a loss and that of the insured who do suffer a loss. At the threshold number of winning risks where the insurer breaks even and for more wins, each insured also receives a refund. For fewer wins than that threshold, each insured either has a losing risk, in which case the compensation is only partial, or has a winning risk, in which case no refund occurs. To increase realism, consider that insolvency involves some cost of administering the insolvent insurer. The coverage of the losing risks is correspondingly smaller.

A preliminary choice is the level of contribution by the insured. That contribution determines the number of winning risks where the insurer breaks even. If the group only wants to face an insolvent insurer with 1% probability, for example, they should set the contribution so that the contributed amounts are sufficient at the number of wins that occurs with probability 1% or more, i.e., for few wins and many losses. The cumulative distribution function for the hundred risks answers that question. As we saw in the example of the cointosses, the appropriate distribution is the binomial, making that problem tractable. Where does the cumulative distribution function of the binomial distribution for a hundred risks, each of which has a 50% probability, reach the corresponding value, in this example the value of 1%?[1] It turns out that

---

[1]In mathematical notation, we seek the greatest $x$ for which

CDF(BinomialDistribution(100, .5), x)$\leq$.01

Because Mathematica cannot solve this, create a loop that obtains the answer by decrementing a variable ($x$) from the maximum number of risks (and where the CDF is 1) until the value of the CDF falls below the target (here .01):

```
x=100;
While[CDF[BinomialDistribution[100,.5],x]>.01,x=x-1]
```

The outcome of the loop is that $x$ is the largest number of wins where the cumulative probability is at or under 1%. To decrement $x$, instead of $x=x-1$ we can write $x--$ .

37 wins (and 63 losses) are the largest number of wins that occur with probability under 1%.

For the mutual insurer to be able to cover the loss of 63 risks, the insurer must raise $63 in premiums from the 100 insured, or $0.63 from each insured.

Once the insurer has been set up, the risks will materialize and the wins can range from none to 100 (and the losses, conversely, will range from 100 to zero). For each possible number of wins, (the probability of which comes from the PDF of the binomial distribution) obtain the expected utility of the insured. This depends on whether each insured is a winner or a loser of the risk and whether the insurer is solvent or insolvent.

If the insurer becomes insolvent, in this example, if the wins are fewer than 37, then the insurer will not have enough assets to compensate all the insured who suffered harm. Rather, the insurer's assets will be divided *pro rata* among the insured. Thus, for wins $w < 37$, each losing insured will suffer harm of $1 minus the assets of the insurer divided by the number of losses, in this example $63/(100 - w)$. Those insured with the losing risks will have also paid the insurance premium, so their final wealth is their initial wealth of $3, minus the uncompensated harm, mitigated by the distribution of assets, also minus the payment to the insurer.

If the insurer becomes insolvent, the winners enjoy the gain. However the winners also have paid the premium. Therefore their terminal wealth is their initial wealth plus the winning of $1 minus the premium of $0.63.

If the insurer remains solvent, which in this example happens if the insured experience more than 37 wins, then the insurer makes whole the losing insured and distributes its remaining assets akin to a refund or liquidating dividend. The remaining assets are the amount raised, in this example $63, minus the coverage paid to those with losing risks, depending on the number of wins $w$, $100 - w$. Accordingly, the insured with losing risks have terminal wealth equal to their initial wealth minus the premium plus the distribution. The loss is covered by the insurer. The insured with winning risks have terminal wealth equal to their initial wealth plus the winning risk minus the premium plus the distribution. The full code of this model is in the appendix.

Figure 2 holds this illustration. Risk aversion ranges from zero at left to about 30 on the right and the insurer's solvency ranges from virtually zero probability of insolvency near the viewer to always being insolvent at the far end, which is equivalent to no insurance. The height of the solid is the change in the certainty equivalent wealth, subtracting the expected utility of wealth with insurance from that without insurance. In other words, the vertical axis holds the gain due to insurance in terms of expected utility of wealth.

Consider first the example of zero risk aversion, i.e., risk neutrality, the heavy line on the left end of the surface that drops below the horizon. Individuals are risk neutral, they treat the expected gains and losses at their probabilistic averages. Insurance does not produce a gain. However, the cost of insolvency becomes a loss. Accordingly, the risk-neutral dislike settings where the insurer may become insolvent and prefer no insurance. However, they become indifferent between no insurance (100% probability of insolvency) and a very safe insurer. The corresponding line, dips immediately as insolvency drops below 100% at the far left end of the graph and only reaches the same level of welfare when insolvency becomes extremely unlikely at the near left end of the graph.

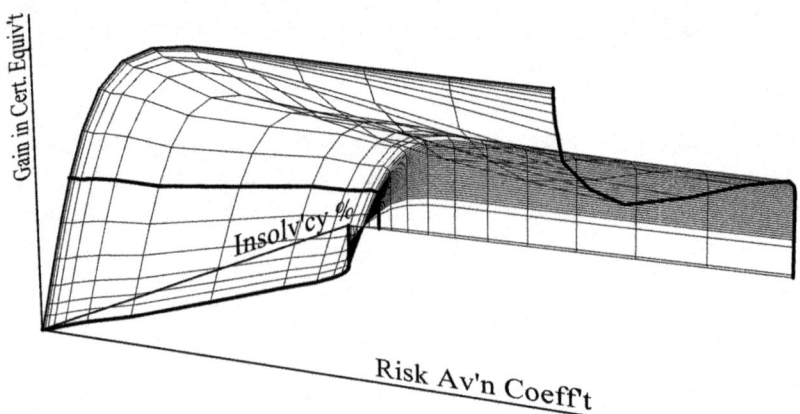

**Fig. 2**  Gain from insurance

At the opposite extreme, consider the individuals with the strongest aversion to risk, the heavy line at the right end of the surface. Any insurance is a significant gain, despite the costs of insolvency. At the opposite extreme, as very low values of probability of insolvency obtain, near the viewer, an additional gain materializes, the avoidance of the cost of insolvency. The corresponding line rises sharply as an insurer comes into effect, however likely to be insolvent, at the far right of the graph. As that line approaches the elimination of the likelihood of insolvency (near the viewer), expected utility rises again sharply corresponding to the gain from avoiding the costs of insolvency. The middle values, however, are relatively flat. Whereas when the insurer was grossly underfunded, the little effect of insurance was desirable, when the insurer is better funded but still likely to become insolvent, the insured with this strong of an aversion to risk experience little gain. The welfare gain from better insurance due to reduced probability of insolvency almost cancels out with the welfare loss due to the increased costs of insolvency. This is due to the fact that the insurer becomes better funded near the viewer, but the costs of insolvency are a percentage of its assets.

The heavy line at about a fifth on the left corresponds to a risk aversion coefficient of about two. This is a less remarkable line. The insured experience a sharp gain from even an undercapitalized insurer and continue experiencing gains gradually as the insurer's insolvency becomes more unlikely near the viewer.

Figure 3 presents only those three lines instead of the surface formed from considering intervening values of risk aversion. The horizontal axis holds the probability of insolvency of the insurer. When it reaches certain insolvency, a value of 1, no insurance exists. The vertical axis holds the change in certainty equivalent from the adoption of insurance for the risk-averse individuals.

The point of the figures is that insurance constitutes a very large gain in welfare. The figures expose individuals with wealth of 3 to a 50-50 risk of losing or gaining 1, a third of their wealth. For the more averse to risk, insurance increases the certainty-equivalent wealth by a sixth of their wealth, provided the insurer stays solvent.

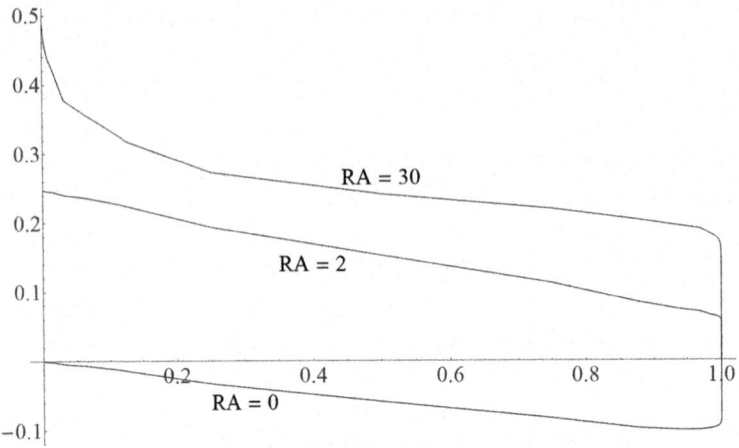

**Fig. 3**   The effect of insurance, three levels of aversion to risk

# 5    Estimates of the Risk-Aversion Coefficient

Numerous scholars have attempted to estimate individuals' coefficient of risk aversion. The resulting estimates, however, have differences that are so large as to seem incompatible. For example, a study of financial instruments compared their risks and returns and produced estimates of about two.[2] While this seems plausible, many of the participants in the markets are financial institutions rather than individuals; assuming the institutions are owned by diversified investors implies that they should be risk-neutral, with the further implication that these are likely underestimates. A study of individual gamblers and option traders suggests

---

[2]R. Mehra & E.C. Prescott, *The Equity Premium: A Puzzle*, 15 J. MONETARY ECON. 145, 154 (1985) (collecting estimates from various scholars mostly from 1 to 2 and positing that its value should be under ten). Others, focusing on the pricing of agricultural commodities produce extremely low estimates of risk aversion; *see generally* Rob Raskin & Mark J. Cochran, *Interpretations and Transformations of Scale for the Pratt-Arrow Absolute Risk Aversion Coefficient: Implications for Generalized Stochastic Dominance*, 11 W.J. AGRIC. ECON. 204, 205 & Table 1 (1986) (collecting estimates, all near zero).

values for the coefficient of risk aversion of 30 and above.[3] Values higher than ten also seem consistent with the previous subchapter on insurance. The idea that an insurer can become insolvent is very strongly deterred by regulation and the media. This suggests that society does not behave according to the implications of risk aversion coefficients under ten, when the insurer's insolvency is not strongly undesirable.

## 6    Deriving the Utility of Wealth Function

So far, the discussion did not look inside the utility of wealth function. The method of differential equations allowed mathematicians to derive their specifics. To switch to differential equations, attitudes about risk need to be restated using derivatives of the utility of wealth. Essentially, using as a foundation Fig. 1, let the fraction of wealth at risk, $f$, approach zero and think in slopes rather than points. One might focus on comparing the slope immediately before to that immediately after the expected wealth $w$. However, that is the definition of how slope changes, i.e., the second derivative of the utility of wealth function. While it is tempting to say that if that slope changes little, then risk aversion is weak, whereas if the second derivative is large, then aversion to risk is strong, the accurate statement is that the rate of change of the slope corresponds to aversion to risk. The second derivative measures change of the slope but the rate of change of the slope is the ratio of the second to the first derivative. Accordingly, mathematicians could write $-u''(w)/u'(w) = a$, but this corresponds to constant absolute risk aversion because aversion to risk, $a$, does not depend on wealth. Also, the second derivative takes a negative sign because it is decreasing, so it is negative, but we seek a measure that would be increasing in aversion to risk. By having that negative sign, the coefficient of risk aversion becomes positive and increasing as aversion to risk increases. To move

[3]Karel Janecek, *What Is a Realistic Aversion to Risk for Real-World Individual Investors?* (2004), www.sba21.com/personal/RiskAversion.pdf [perma.cc/S86X-YD8N]. *See also* S. Kandel & R.F. Stambaugh, *Asset Returns and Intertemporal Preference*, 27 J. MONETARY ECON. 39 (1991) (also allowing the plausibility of a value of 30 for the coefficient of risk aversion).

from absolute to relative aversion to risk, make this depend on wealth. While we could multiply $a$ by wealth, the convention is to divide it by wealth or multiply the left-hand side by wealth. Thus, mathematicians write $u''(w)/u'(w) = -a/w$, calling $a$ the coefficient of risk aversion. The equation is usually rearranged as $-w\,u''(w)/u'(w) = a$.[4] The resulting utility of wealth that displays constant relative risk aversion is $u(w) = \frac{w^{1-a}}{1-a}$.

This general form of the utility of wealth function has the problem that it cannot handle the case of $a = 1$. Therefore, the case of the coefficient of risk aversion taking the value of one is a special case that needs to be solved separately.[5] This turns out to be the natural logarithm of wealth: $u(w) = ln(w)$.

Figure 1 uses this constant relative risk aversion function for the utility of wealth with a coefficient of risk aversion of $a = 4.5$. Exercise 4 has you produce an animated version of the graph that lets the user change risk aversion with a slider and observe the change in the curvature of the utility of wealth function.

Let us verify the math with some examples. Take an individual with constant relative risk aversion and coefficient of risk aversion $a = 2$. This individual should react the same way to the following two gambles that are same fractions of two different levels of wealth. Case 1 has our individual having wealth of \$1000 and facing a 50-50 chance of winning or losing \$100. Case 2 has our individual having wealth of \$10,000 and facing a 50-50 chance of winning or losing \$1000. What is the certainty equivalent wealth of this individual in each case?

---

[4]We can use Mathematica to solve this differential equation. The result is $u(w) = \frac{w^{1-a}C_1}{1-a} + C_2$. The differential equation approach gave us a function of wealth $w$ and risk aversion, $a$, but two additional parameters appeared, $C_1$ and $C_2$. Those are constants of integration. Solving differential equations without having specified where to place the resulting curves on the plane leaves constants of integration. Thinking back to the curve of Fig. 1, changing $C_2$ moves that curve in the vertical direction and changing $C_1$ moves it in the horizontal direction. If we had a specific value of utility that some amount of wealth should create, then the constants of integration would make the resulting function comply with this specification. However, because utility of wealth is not measurable, we do not have any such specification. Rather, the convention is to generalize away the constants of integration. They disappear by taking neutral values, zero for $C_2$ and one for $C_1$.

[5]The raw solution is $u(w) = C_1 log(w) + C_2$, where $log()$ is the natural logarithm, often noted as $ln$. Replacing $C_1$ with 1 and $C_2$ with 0 gives the clean result in the text.

# 7    Legal Change

A perspective comparing the common law and the civil law legal environments, as mechanisms of implementing legal change with different frequencies, relates them back to aversion to risk.[6] Consider that legislative change can more easily be large whereas change by a court's new interpretation of existing law tends to be more incremental. Legislatures also face costs to act in the sense that collective decisions can be impractical. As a result, legislatures tend to only act if the needed change is significant. Public choice arguments about the difficulty and biases of collective action would also point toward the rarity of legislative action compared to judicial interpretation, which occurs as a natural byproduct of adjudication.

A legal system needs to update itself to changing socioeconomic circumstances. The legal system can choose the updating method of the common law or that of civil law. Common law allows nearly constant but gradual change by the courts. Civil law means that change comes from the legislature in less frequent but larger increments.

Individuals subject to legal change face a risk. The change may be advantageous or not. Granted, if socioeconomic pressure exists for a change, this suggests that changes will tend to be desirable. Specific individuals, however, can be on the losing side of changes. Accordingly, large changes bring greater fear than small changes. It stands to reason, then, that as aversion to risk increases, individuals will tend to prefer systems of frequent gradual change, common law systems, over systems of less frequent, larger change, civil law systems.

# 8    Conclusion

Despite that the mathematical features of the analysis of aversion to risk are complex, the topic is central to the making of decisions. Many innovations in contracting, such as insurance, and in law, such as the limited

---

[6]A fuller analysis appears in my article Nicholas L. Georgakopoulos, *Predictability and Legal Evolution*, 17 INT'L REV. L. & ECON. 475–89 (1997).

liability of corporations, seek to address aversion to risk and produce economic growth closer to that of a risk-neutral society. Even the success of the common law (and generally a more frequent but gradual change of law and institutions) may be attributable to risk aversion. The loss aversion that behavioral studies find, may well be a mere side-effect of risk aversion.

# 9     Exercises

1. Abi and Bon face the same risk, a 50-50 gamble to win $1000 or nothing. Abi is willing to sell the position for $450 but not less. Bon is willing to sell the position for $400 but not less. (a) Who is more averse to this risk? (b) What are the $450 and $400 figures called?
2. Use Mathematica's **DSolve[    ]** function to find the utility-of-wealth function that produces constant relative risk aversion (a) for a variable coefficient of risk aversion $a$; and (b) for a coefficient of risk aversion of one ($a = 1$).
3. Plot the function by selecting a range of values for wealth and plotting the above function for your chosen value of the coefficient of risk aversion.
   3.1 Select the fraction of wealth that is subject to a 50-50 gamble and add to the plot the points that correspond to a loss and a win (hint, you could place the Point[ ] commands inside an Epilog−>{ } option).
   3.2 Add to the plot a line connecting the above points.
   3.3 Add to the plot a point corresponding to the utility of having the average wealth.
   3.4 Add to the plot a point corresponding to the average utility expected from the gamble.
   3.5 Reposition the axes to the values that seem appealing, deriving those values from the values that the utility function takes and the difference between the utility of winning and losing the gamble.
   3.6 Solve the certainty equivalent.

3.7 Add trace lines and labels using the derived values.

3.8 Restate the plot range from the values the function takes.

3.9 Test that the graph is pleasing for low values and high values of risk aversion (such as $a = .5$ and $a = 15$). The result should be similar to Fig. 1.

4. Paste your code inside a **Manipulate[ ]** command, deleting the fixed specification of the coefficient of risk aversion and specifying a range for it as part of that command, perhaps from zero to fifteen. Separate the commands with commas. Ensure that a correct result for $a = 1$ appears (hint: you could use the **If[ ]** function). The result should produce a version of Fig. 1 where the user changes risk aversion and observes the change in the curvature of the utility-of-wealth function.

5. Consider the example of the risk-averse ranchers bargaining with farmers about crossing the farm instead of taking the long cattle drive, on page 8 in the chapter discussing Coasean irrelevance and its failure due to aversion toward risk. Suppose the wealth of the rancher is $200. The rancher faces a 50-50 chance that the cattle on the long drive will misbehave. If they do not misbehave, then the long drive only costs $5. If the cattle misbehave, then the long cattle drive becomes much more involved, requiring extra effort and some losses and costs $95. Although a risk-neutral rancher would see this as a $50 cost, the rancher considers that the certainty equivalent of the long cattle drive is a cost of $80. The rancher's utility-of-wealth function exhibits constant relative risk aversion. What is the rancher's coefficient of risk aversion?

## Appendix: Code for Figure 2

```
(*loop of insolvency values of the insurer*)
Clear[x, y]
y = 1;
array1 = Table[y = y/2, {150}];
array2 = 1 - array1;
```

```
loopsol =
  Flatten@{Reverse@array1, Drop[array2, 1], 1(*drop the
    duplicate 1/2, add a final 1*)};
(*loop of values of risk aversion*)
Clear[x, y, ex]
ex = 4;
looprac =
 Flatten@{0,
    Table[x^ex , {x, .2^(1/ex), 32^(1/ex), (32^(1/ex) - .25^(1/
ex))/
     15}]};
mintarg = 0;(*solvency loop minimum*)
maxtarg = .9999999999;(*solvency loop maximum*)
ni = 100;(*number of insured*)
ps = .5;(*probability of success of each insured, i.e., prob-
ability of the gamble winning, of not suffering harm but hav-
ing gain*)
(*set premium so that the insurer becomes insol-
vent with only target (say 1%) probabil-
ity: for how many wins is the CDF[Binomial[ ],wins] equal to
target?*)
target =.; cra =.;
w0 = 3;(*initial wealth of each insured*)
wg = 1;(*size of wager of each insured in the double or noth-
ing risk/venture*)
ci = .25;(*cost of insolvency, fraction lost to law-
yers and auctioneers*)
surfc = Table[{
 x = ni;
 While[CDF[BinomialDistribution[ni, ps], x] > target, x-];
 pm = (wg (ni - x))/ni;(*premium payment by each insured*)
  (*Print["The probability of failure is under ",target
100,"% at ",
 x," wins, at most. The payment is ",pm//N,"."]*)
 wt[target] = x;(*store the breakeven wins*)
 x =.;
 mibi = ni - (ni pm)/wg;(*equal to wt[target];
 number of wins making the mutual insurance barely insolvent,
 result of Solve[wg (ni-x)\[Equal]ni pm,x], i.e.,
```

how many wins make the lost wagers equal to the assets of t
he insurer?*)

(*if fewer than mibi wins,

then the losers suffer uncompensated harm.

In these cases we need the amount of the uncompensated harm. The assets of the mutual insurer, ni pm, reduced
by the cost of insolvency ci to (1-ci)ni pm, are divided
proportionately among the number of actual non-wins, ni-wins;
therefore instead of losing wg they lose wd minus the distributed assets.*)

uh = wg - ((1 - ci) ni pm)/(ni - wins);(*uncompensatd harm.
the lost wager reduced by the distributed assets,
which are the assets of the insurer ni pm divided by the losse
s, ni-wins; Verification: should be zero when winsmibi except
in the case of no insurance, target=1, when it will be 1.*)

lii = w0 - uh - pm;(*wealth of losers if the insurer becomes insolvent*)

(*the rest of the time, with probability wins/ni,
the insured wins and ends up with wealth w0+wg-pm*)

wii = w0 + wg - pm;(*winning insured wealth if the insurer
becomes insolvent*)

wmni = (ni pm - (ni - wins) wg)/ni;(*per insured wealth of
mutual insurer if it does not become insolvent, after if
compensates all harmed insureds, distributed to the ni members. Becomes zero if the number of wins is mibi*)

wis = w0 + wg - pm + wmni;(*wealth of winners if the insurer stays solvent, i.e., for wins>mibi*)

lis = w0 - pm + wmni;(*wealth of losers if the insurer stays sol-
vent, i.e., for wins>mibi; note: the insurance is full, compensating their entire loss, else we would need to subtract a deductible*)

eu = Sum[
    PDF[BinomialDistribution[ni, ps], wins] If[
        wins <= wt[target] (*mibi*),
        (ni - wins)/ni u[lii, cra] +
        wins/ni u[wii, cra], (ni - wins)/ni u[lis, cra] +
        wins/ni u[wis, cra]],

```
    {wins, 0, ni}]; (*expected utility: proba-
bilty of being a loser or a winner, times the resulting util-
ity of wealth*)
  (*corrsponding certainty-equivalent wealth*)
  Off[Solve::ifun];
  ceiA =
   x /. (Solve[u1[x, a] == eu, x] //
      Simplify)[[1]]; (*certainty equivalent*)
  cei1 =
   x /. (Solve[u2[x] == eu, x] //
      Simplify)[[1]]; (*certainty equivalent if a=1*)
  twi = If[cra == 1, cei1,
    ceiA /. a -> cra]; (*get certainty equivalent*)
  (* get certainty equivalent with no insurance:*)
  cea = x /. (Solve[
      u1[x, a] == (1 - ps) u1[w0 - wg, a] + ps u1[w0 + wg
, a], x] //
      Simplify)[[1]];
  cel = x /. (Solve[u2[x] == (1 - ps) u2[w0 - wg] + ps u2[w0
+ wg],
      x] // Simplify)[[1]];
  On[Solve::ifun];
  tw = If[cra == 1, cel,
    cea /. a -> cra]; (*get certainty equivalent*)
  cra, target, twi - tw},
  {cra, looprac}, {target, loopsol}];
mincex = 0; maxcex = 32; mincey = 0; maxcey = max-
targ; mincez = 0; \
maxcez = .55;
inswelfeff1 = Graphics3D[{Black,
  Line /@ surfc,
  Line /@ Transpose@surfc,
(*labels*)
    Text["Risk Av'n Coeff't", {.82 max-
cex, mincey, mincez + .025}],
  Text["Insolv'cy %", {mincex, .8 maxtarg, mincez + .03}],
  Text["Gain in Cert. Equiv't", {0, -.03, .75 maxcez}],
  Thickness[.002], (*axes*)
  Line[{{mincex, mincey, mincez}, {maxcex, mincey, mincez}}],
  Line[{{mincex, mincey, mincez}, {mincex, maxcey, mincez}}],
```

```
Line[{{mincex, mincey, mincez}, {mincex, mincey, maxcez}}],
Thickness[.004],
Line[surfc[[1]]],
Line[surfc[[7]]],
Line[surfc[[-1]]]
},
BoxRatios -> {1, 1, .6}, Boxed -> False,
ImageSize -> 7 imgsz,
BaseStyle -> fntandsz,
ViewPoint -> {1.75, -3, .65}]
```

# 11

# Financial Crisis Contagion

## 1    Introduction: Fear Crises!

A book about finance must discuss financial crises. Whereas the discussion of finance in the rest of this book indicates rationality and, by implication, stability, the lived experience of finance is a sequence of crises—enthusiasms or exuberances followed by panics. Psychology interacts with the interconnectedness of financial institutions to produce the rare nightmare of financial crises. Despite our understanding of both psychology and finance, a solid fear of the next financial crisis is the most healthy attitude.

Small financial crises happen with some regularity every ten to fifteen years. Often the cycle of exuberance and panic corresponds to unusual economic growth of a sector and its end. When the markets expect growth in some class of assets, the markets will value those assets by discounting to the present the expected future growth. The rosy outlook for this asset class almost inescapably leads to borrowing that uses the asset as collateral. One day, the expectation for continued growth is dispelled. The prices of the assets drop suddenly and significantly because the future growth in values is no longer true. The individuals

© The Author(s) 2018                                                           **201**
N. L. Georgakopoulos, *Illustrating Finance Policy with* Mathematica,
*Quantitative Perspectives on Behavioral Economics and Finance,*
https://doi.org/10.1007/978-3-319-95372-4_11

and businesses that borrowed and the financial institutions that lent to them find themselves in trouble. A fraction of borrowers and a fraction of lenders become insolvent. They sell collateral, the previously prized assets, voluntarily and involuntarily pursuant to foreclosures. This further depresses the prices of the previously prized asset class, further aggravating the insolvencies of borrowers and lenders. In 1929, the prized asset was stocks. In 1989, the prized asset was real estate. In 1998, the prized asset was internet stocks (naming it the Dotcom bubble). In 2008, the prized asset was again real estate, in large part transformed into mortgage-backed securities.

When the crash is not large, then its consequences stop at these failures of borrowers and lenders and a temporary and small reduction in the availability of credit. The result is not devastating for the economy. The availability of credit drops not only because the failed lenders cease operations, but also because the reduced value of the assets makes most economic actors need more cash, reducing the lending capacity of even the lenders who remain solvent. When the crash is large, however, not only are the corresponding failures of borrowers and lenders a greater fraction of the economy, but also the ensuing reduction in the availability of credit is large and becomes devastating, forcing any business that is short of cash to become insolvent. The result is a destruction of much of economic capacity, as the Great Depression demonstrated. In 2008, the events were poised to lead to an even more devastating depression.[1] Very aggressive action by the regulators stopped the cascade. The most visible measures were dramatic. (1) The Federal Reserve made loans to banks and nonbank financial institutions. (2) The Federal Deposit Insurance Corporation guaranteed all obligations of all banks. (3) The Federal Reserve made loans (by buying commercial paper) directly to businesses. (4) Regulators strong-armed banks into accepting an infusion of funds.[2] Because the Federal Reserve no longer has the legal

---

[1] HENRY M. PAULSON Jr., FIVE YEARS LATER: ON THE BRINK—THE NEW PROLOGUE: A LOOK BACK FIVE YEARS LATER ON WHAT HAPPENED, WHY IT DID, AND COULD IT HAPPEN AGAIN? 40 (2013) ("Unemployment levels that topped out at 10 percent ... could easily have risen to 25 percent").

[2] See generally BEN S. BERNANKE, THE FEDERAL RESERVE AND THE FINANCIAL CRISIS 78 (Princeton Univ. Press ed., 2013) ("The Fed provided cash or short-term lending to [brokerage houses] ... Commercial paper borrowers received assistance, as did money market funds... [T]he Fed created

authority for as drastic (and as speedy) an effort to resuscitate the economy, a fear that a subsequent large financial crisis will run amok is well founded.

Moreover, a depression can have much worse consequences than merely economic; depressions influence political decisions. The evidence shows that voters react to recessions by (a) moving away from centrist positions to more extreme positions, somewhat more to the political right than the left, and (b) turning against immigrants and minorities.[3] The rise of the Nazi party to power in Germany in the early 1930s has been linked to the depression.[4] Recognizing that the rise of the Nazi party was a consequence of the depression shows that both the most destructive war in history, WWII, and the Holocaust of millions of Jews and other minority groups was an indirect result of the mishandling of the financial crisis of 1929 and the Great Depression.

We must fear financial crises. Always.

# 2    Finance in the Cycle of Production

To observe how financial crises become economic depressions, the first step is to recognize that economic activity is a cycle. Businesses employ individuals to provide products and services. Individuals spend their

---

some new liquidity programs to help get [the asset backed securities market] started again, which we were successful in doing."); HAL S. SCOTT, CONNECTEDNESS AND CONTAGION: PROTECTING THE FINANCIAL SYSTEM FROM PANICS 75–78 and Table 7.2 (2016) (listing the responses after the run on money market funds).

[3]Manuel Funke, Moritz Schularick & Christoph Trebesch, *Going to Extremes: Politics After Financial Crises, 1870–2014*, 88 EUROPEAN ECON. REV. 227 (2016) (finding recessions lead voters toward the extremes and toward anti-immigration and anti-minority parties). *See also* MASS POLITICS IN TOUGH TIMES: OPINIONS, VOTES AND PROTEST IN THE GREAT RECESSION (Nancy Bermeo & Larry M. Bartels, eds., Oxford U.P. 2014) (explaining why voters did not move to extremes more); Larry M. Bartels, *Political Effects of the Great Recession*, 650 ANNALS AM. ACADEMY POL. AND SOC. SCI. 47, 48, 68 (Nov. 2013) (post-recession rise of right-wing populism, plus cultural and racial tensions).

[4]*See, e.g.*, Alan de Bromhead, Barry Eichengreen, & Kevin H. O'Rourke, *Political Extremism in the 1920s and 1930s: Do German Lessons Generalize?*, 73 J. OF ECON. HIST. 371 (2013) ("confirm[ing] the existence of a link between political extremism and economic hard times"). *See generally* IAN KERSHAW, HITLER: 1889–1936 HUBRIS (2000).

earnings on products and services, completing the cycle. What individuals do not spend, they save, providing those funds, directly or indirectly, to businesses as capital. Businesses, in turn, pay a return on that capital. From this perspective, economic activity forms two concentric cycles, the cycle of production and consumption, which is the bulk of economic activity, and the cycle of investment and its returns.

Figure 1 illustrates this vision of the economy as two concentric cycles. Individuals are on the right and businesses are on the left. The outside cycle corresponds to production and consumption and is heavier than the inside cycle that corresponds to investment and its returns. Money travels counterclockwise in both cycles. In the outer cycle, on production and consumption, individuals spend on consumption, businesses receive this amount as their sales. Businesses, in turn, employ individuals and spend on payroll, which individuals receive as wages. In the inner cycle, on investment and its returns, fewer individuals invest in businesses what from the perspective of the businesses is capital. The businesses turn and pay their investors a return which from the perspective of the business is the cost of their capital.

If the money (and, therefore, the corresponding products and services) travels faster, then the economy is more productive. Businesses

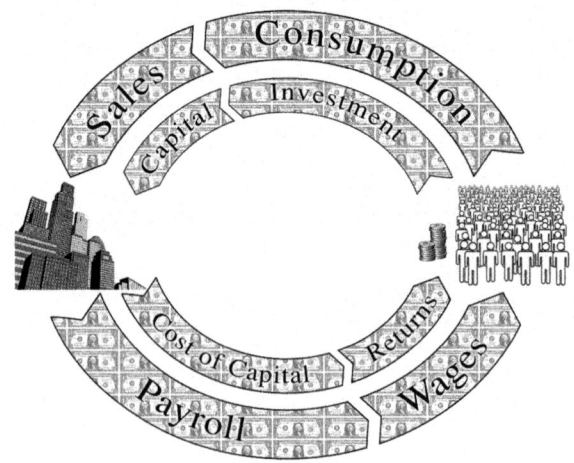

**Fig. 1** The interdependence of the real sector with the financial

sell more and pay more to their employees and investors, who enjoy more products and services. Vice versa, if the flow of money slows, then the economy becomes less productive. Businesses earn less and can pay less to employees and investors, who, in turn, get fewer products and services.

This cycle of economic activity has been analogized to a spinning top. Macroeconomic scholarship and the work of the central bank, essentially, focus on how to prevent the spinning top from slowing down and how to keep it accellerating smoothly. Excess acceleration may produce inflation. Slowdowns are recessions or depressions. The objective of this chapter is to discuss how the regulation of financial institutions tries to prevent a financial crash from turning into a recession or depression for the economy.

Neither the cycle of production and consumption, nor the cycle of investment and its returns, are impossible without financial institutions. With some stretching of the imagination, one can envision both cycles operating without financial institutions. Employees could be paid in goods and in service vouchers, which they could barter for the goods and services they would consume or the goods that they would rent to businesses as their investments. However, that would be a nightmarishly complex economic system. By the interjection of finance, employees earn money which they spend directly on consumption or investments.

Financial institutions greatly facilitate both cycles. Financial institutions help (1) by facilitating the movement of value; (2) by smoothing over situations when a few economic actors' income temporarily does not cover expenses; and (3) by the central bank's powers to counteract recessions. The facilitation of the movement of value consists of financial institutions helping make faster the transfer of value from one economic actor to the next. Compare, for example, having to mail a crop of fruit to a distant juicing facility to transfering the funds to buy the same amount of fruit. The second function consists of lending to the economic actors who temporarily cannot meet their obligations. Compare a world where the business that has a temporary shortage of sales is forced to delay paying its landlords, suppliers, and employees to one where it can borrow and pay them on time. In the former setting more businesses will close while being viable (because landlords will evict and employees will quit) and more individuals will forego consumption that they can afford. Financial institutions prevent the false closures and the reduction of consumption by lending to the viable businesses and consumers. The third function, on the surface, consists of the central bank's changing interest rates and the quantity of money in the economy to counteract euphorias and panics.

The function of the central bank becomes vastly more important during a large crisis. Even in a small financial crisis, some financial institutions fail. Nevertheless, the disruption of the financing of the cycle of economic activity is not large. Granted, some viable businesses close because they cannot get loans; some individuals cannot engage in consumption that they could have afforded; some individuals lose employment due to the closure of businesses. Nevertheless, in a small crisis, the low interest rates and additional money that the central bank provides are enough to return the economy to its prior productivity.

In a large crisis, however, the lending by financial institutions drops so profoundly that such a large number of businesses fail and so many individuals are in dire straits, that, in turn, the lower rates and the additional money supplied by the central bank do not suffice to bring the economy back to its prior level of productivity. This is the kind of crisis that occurred in 2008. The central bank flooded the market with money and reduced rates to almost zero, but it also had to lend directly to businesses in order to prevent massive businesses closures and to keep the economy going.

The legal system addresses crises in many ways, some intentional and some, likely, unintentional; some only tangentially related to crises and some directly addressing crises.

# 3    Policies About Financial Crises

The discussion of the policies that address financial crises proceeds in four steps. The first subsection discusses the policies that are tangentially related to financial crises, meaning that these policies have other, everyday concerns as well, but they play an important role during financial crises. The second subsection discusses policies that primarily address financial crises. The third subsection explains how all these measures failed in 2008 and how the Federal Reserve managed to have the economy escape a new great depression using extraordinary measures. Finally, the fourth subsection discusses the policy reactions to the financial crisis.

## 3.1 Policies Indirectly Related to Crises

Bankruptcy law, the social safety net, and the regulation of securities markets do not aim directly at crises but take special importance during a financial crisis.

### 3.1.1 Bankruptcy and Reorganization Law

Bankruptcy law is not designed for crises but it becomes especially important during crises. Businesses that become insolvent during times of general economic health, pose small problems and tend not to destroy productive capacity of the economic system. If the assets of the business would be more productive at a different line of business, then the failed business will tend to be liquidated. The assets of the business will be sold and redirected to more productive use. If the assets of the business are more productive in their current use, i.e., configured as this business, then the business will either be sold as a going concern or will undergo a reorganization inside bankruptcy.[5] The healthy economic environment facilitates the sales. Buyers forecast growth rather than contraction and lenders are willing to finance sound purchases. At times of economic health, one could even go as far as to say that the absence

---

[5]One may wonder how such a business may become insolvent at times of economic health but the paths to insolvency are many. Perhaps an accident occurred that reduced the business's output or imposed an unmanageable obligation. For example, a facility might succumb to an uninsured calamity or a massive tort liability might hit the business. Then, the income stream of the business cannot service its obligations and insolvency arises despite the productivity of the business and the absence of a financial crisis.

of the reorganization chapter from the bankruptcy code would not drag much on economic productivity.[6]

During a financial crisis, by contrast, the bankruptcy system faces difficult decisions and plays an important role in preserving productivity for two reasons. First, buyers and lenders will not serve the function of separating the businesses that must stay intact from those that must be liquidated. Second, the dearth of buyers and lenders means that reorganizations must occur.

First, whereas at times of general economic health, buyers and lenders separate the businesses that can remain intact from those that should be broken apart, at times of crisis buyers and lenders are not as forthcoming. Buyers foresee contraction and tend not to make purchases. Lenders face the problems that the crisis creates, are in a credit crunch, and have a low willingness to lend. The result is that the task of deciding which businesses should remain intact falls on the bankruptcy courts and more errors tend to occur. Some businesses that should be liquidated will continue to waste resources in reorganization but, even worse, some businesses that should be reorganized will be liquidated, destroying productive capacity.

Second, whereas during times of general economic health, insolvent businesses that ought to stay intact as going concerns can relatively easily be sold to willing buyers who are financed by willing lenders, at times of crisis the scarcity of buyers and lenders means that such businesses must be reorganized. Reorganization law, effectively, imposes a voluntary (the contradiction between *imposes* and *voluntary* is intentional) renegotiation of the obligations and the ownership structure of

---

[6]Indeed, economic systems have performed acceptably well without reorganization law. The primary example in the United States itself before the addition of the reorganization chapter to the old Bankruptcy Act in 1933. Railroads would become insolvent and would get reorganized through foreclosure sales in equity receivership to entities formed by their creditors. Professor Per Strömberg finds the performance of a system without reorganization law in Sweden satisfactory; *see* Per Strömberg, *Conflicts of Interest and Market Illiquidity in Bankruptcy Auctions: Theory and Tests*, 55 J. Fin. 2641 (2000). For a thorough assessment of the new reorganization chapters in the UK, Germany and Sweden, *see* David Smith & Per Strömberg, *Maximizing the Value of Distressed Assets: Bankruptcy Law and the Efficient Reorganization of Firms, in* Systemic Financial Crises: Containment and Resolution 232 et seq. (Patrick Honohan & Luc Laeven, eds., 2005).

the failed business, subject to rules that are vague and imperfect. An imperfect reorganization legal system will produce some reduction of economic capacity by liquidating businesses that should have stayed as going concerns and will produce some waste by keeping intact businesses that should be liquidated, tying up capital and lending capacity. Even if the bankruptcy system performed perfectly, liquidating with utmost accuracy only the businesses that should be liquidated, the end result would not be happy lenders. The quasi-forced nature of reorganizations means that lenders who would have preferred to no longer lend to the reorganized businesses end up either owning them or continuing to lend to them; that capital, although not being wasted, is not available to finance more hopeful projects.

### 3.1.2 The Social Safety Net

The social safety net primarily consists of labor and employment law, unemployment benefits, health and disability insurance, retirement benefits and their guarantee by the Pension Benefit Guarantee Corporation (PBGC). The social safety net protects the economically weakest and saves individuals from the worst outcomes of the free market. Thus, included in the safety net should be the efforts to address homelessness and poverty, mental health, as well as the rules about eleemosynary organizations and donations to them.

These areas of law are important even at times of general economic health. At times of crisis they take crucial importance for society's welfare and even for the actual survival of swaths of the population. Nevertheless, these areas of law have a weak productivity-restoring function. Granted, worker retraining programs aim directly to restore the productivity of workers with outdated skills. However, some scholars even argue that the social safety net prolongs the time of low productivity out of the workforce.[7] Proposals to improve the social safety net are useful and important. At times of crisis, the experience is that the social

---

[7] *See, e.g.,* Patricia K. Smith, *Welfare as a Cause of Poverty: A Time Series Analysis*, 75 Pub. Choice 157, 158, 167–68 (1993) (outlining the work disincentive hypothesis and finding support).

safety net is strained. A more flexible organization of the social safety net should absorb some of the additional capacity from unemployment during a crisis into the ranks of the providers of the services of the social safety net, encourage entrepreneurship, and facilitate sorting the supply of labor into the demand.[8]

### 3.1.3 Securities Regulation

The salient parts of securities regulation are (1) the imposition on businesses as issuers of securities of an obligation to disclose (a) accounting information following specified rules on a quarterly (simplified) and annual (detailed) basis; (b) significant news; and (c) trades of insiders; and (2) the creation of a legal regime where sellers (again, businesses as issuers) must disclose all material risks of the securities at the cost of rescission of the sale and liability for the omission. The result of this web of rules is that businesses must follow four policies. First, businesses must disclose every conceivable risk meticulously when issuing securities. Second, businesses must follow standard accounting rules in announcing their numbers (revenues and earnings; assets and liabilities). Third, businesses must announce their numbers quarterly. Fourth, businesses must announce significant news and trades of insiders in their stock.

The direct consequences of the regime of securities regulation are the promotion of accurate prices on the stock market and of active trading.[9] However, these policies secondarily also serve a minor role in averting the exuberances and panics of crises. The airing of everything that can

---

[8]*See, e.g.,* Yoonyoung Cho, David Robalino & Samantha Watson, *Supporting Self-Employment and Small-Scale Entrepreneurship: Potential Programs to Improve Livelihoods for Vulnerable Workers,* 5 IZA J. LAB. POL. 1, 13, 23 (stating that programs that help the self-employed and small-scale entrepreneurs—the two most common types of workers—will help improve livelihoods); *see also* INDEP. EVALUATION GROUP, SOCIAL SAFETY NETS: AN EVALUATION OF WORLD BANK SUPPORT, 2000–2010, 79 (2010), https://openknowledge.worldbank.org/bitstream/handle/10986/21337/672860WP-0Box3600ssn0full0evaluation.pdf [perma.cc/Y6LW-PL84] ("Continued effort is needed to develop [social safety nets] that are flexible and able to respond to shocks; to build country institutional capacity to address various sources of poverty, risk, and vulnerability…").

[9]I have expanded on how securities law facilitates a virtuous cycle from accurate prices to cheaper trading and greater market liquidity, which begets accurate prices; *see* NICHOLAS L. GEORGAKOPOULOS, THE LOGIC OF SECURITIES LAW (2017).

go wrong chills exuberance. Revealing accurately and understandably the performance of businesses and of news dispels rumors and innuendo and tries to preserve rationality to counter both exuberances and panics.

Securities regulation, in other words, adds transparency and tries to add sobriety to the securities markets.

## 3.2    Policies Directly About Crises

The policies that directly address financial crises regard banking. They are (a) the creation of the lender of last resort, the central bank; (b) the take-over and orderly liquidation of failing banks; (c) the insurance of bank deposits; and (d) the rules and supervision that ensure that banks hold sufficient assets that are called capital adequacy rules. The existence of the central bank as a lender of last resort ensures that banks have funds to satisfy withdrawal requests during a crisis. The takeover by the regulators and the orderly (speedy) liquidation of financial institutions ensures their creditors are repaid immediately, preventing the contagion from the wait for the bankruptcy process. The insurance of deposits seeks to prevent depositors from requesting withdrawals during a crisis and to prevent fears about banks' solvency from becoming self-fulfilling prophesies through bank runs. The capital adequacy rules seek to ensure that banks have sufficient assets to meet all their obligations, even during a crisis. These measures, designed mostly in reaction to the Great Depression, had proven sufficient to address all the crises that occurred up to 2007, no small feat. However, in 2008, all these policies proved to be insufficient. The financial crisis produced a sizable contraction in lending, and the regulators had to intervene in radical ways to prevent a new Great Depression. Section 3.3 discusses those extraordinary interventions, whereas this subsection explains the function of these policies.

### 3.2.1 Last-Resort Lending

Banking rules do require that banks hold a sufficient buffer of cash to meet demand. However, in a time of crisis, a buffer of any size can be depleted. Borrowing from the central bank means that the banks which face large withdrawal requests and a drying up of other funding sources can turn to the central bank and obtain funds. This borrowing counteracts bank runs and panicked withholding of financing.

Granted, when the central bank creates additional money, inflation is a concern. However, during a financial crisis the reality is massive deflation. During a crisis, economic actors prize cash because they lose trust in the banks, whereas holding cash at other times tends to be seen as wasteful idleness of capital that could be earning returns. Even if prices do not drop, the panic that drives actors to hold cash, withdraw deposits, and cancel lending, produces an overvaluation of cash and a deflationary pressure. For example, in the Great Recession of 2008, the additional cash that was injected by regulators and the government into the economy was staggering.[10] Nevertheless, the panic was such that, rather than inflation, deflation appeared.[11]

### 3.2.2  Takeover and Orderly Liquidation

The concern about putting banks through the regular bankruptcy process is that it is slow. An ordinary bankruptcy or a reorganization in bankruptcy will take several months. The obligations of the failed bank, by contrast, may be due in days. The concern is that because bankruptcy would only allow the failed bank's creditors to receive value at the end of the process, the creditors, who were due funds from the

---

[10]See generally BEN S. BERNANKE, THE FEDERAL RESERVE AND THE FINANCIAL CRISIS (2013); see also Ivana Kottasova, $9 Trillion and Counting: How Central Banks Are Still Flooding the World with Money, CNN MONEY (Sept. 9, 2016), http://money.cnn.com/2016/09/08/news/economy/central-banks-printed-nine-trillion/index.html [perma.cc/S8VH-6QZ7] (stating that the Federal Reserve injected $3.9 trillion dollars over the course of three rounds of asset buying between November 2008 and October 2014; see also Frederic S. Mishkin, The Financial Crisis and the Federal Reserve, 24 NBER MACROECONOMICS ANNUAL 495, 502 (2009) (explaining the Troubled Asset Relief Program after the Lehman bankruptcy and the AIG bailout).

[11]See Consumer Price Index for All Urban Consumers: All Items, FED. RES. BANK OF ST. LOUIS, https://fred.stlouisfed.org/series/CPIAUCSL (visited Apr. 11, 2018) [perma.cc/DDU2-GUT6] (showing the consumer price index dropping from 218 in July of 2008 to 211 in December of 2008; being flat in 2010; and experiencing occasional small drops in 2012–2017). See also Simon Gilchrist, Raphael Schoenle, Jae W. Sim & Egon Zakrajsek, Inflation Dynamics During the Financial Crisis 1–3 (Fin. and Econ. Discussion Series, Federal Reserve Board, Working Paper No. 2015-012), https://www.federalreserve.gov/econresdata/feds/2015/files/2015012pap.pdf [perma.cc/4LAP-9CT9].

bankruptcy would, in turn, be forced to breach obligations of their own. Thus, even if the creditors would eventually be paid sufficient amounts to stay solvent, the bankruptcy delay would force them into insolvency. The failure of Lehman in 2008 revealed an analogous concern. When Lehman filed for bankruptcy, some money market mutual funds that held claims against Lehman, announced that they would not honor withdrawal requests at 100% pending the bankruptcy because the funds did not know how much they would receive on their Lehman claims. Despite that the bankruptcy estate of Lehman paid these claims eventually almost in full, the market reaction was panic. Institutional investors withdrew massive amounts from money market funds that invested in loans to firms like Lehman.

To avoid the contagion or panic that the delay of a bankruptcy process may cause, bank regulation entitles the FDIC to take over failing banks, i.e., banks that severely violate capital rules. The FDIC takeover accelerates the resolution compared to a bankruptcy process because the debtor, the bank, has much fewer rights than a debtor in bankruptcy has. Rather than allow the debtor to challenge claims and try to reorganize, the FDIC takes the bank over, makes the necessary cash infusion to pay the insured deposits, and determines the payments to which all creditors are entitled.[12] Often, the FDIC will find a different, healthy bank to take over the failed bank. This will often happen over a single weekend, enabling financial life to continue without uncertainties on the open of business the following Monday.

### 3.2.3 Deposit Insurance

Deposit insurance seeks to prevent runs on banks by eliminating the need for a run. Depositors can rest in the comfort that even if their bank were to fail, their deposits would be safe. As a panic produces the threat of insolvency of banks, depositors need not fear for their bank's solvency. In the 2008 crisis, the existence of the limited deposit

---

[12]*See, e.g.*, Golden Pac. Bancorp v. United States, 15 F.3d 1066, 1074 (Fed. Cir. 1994) (rejecting takings arguments against an FDIC takeover).

insurance regime of healthy times opened up an additional possibility. Regulators temporarily authorized the existing mechanism to insure all claims against all financial institutions in order to stem the tide of panic.

Deposit insurance has the negative consequence of reducing the cost of risky behavior (a phenomenon, present in all insurance, called moral hazard). Because depositors' concern about the financial health of their bank is reduced, the bank, in turn, has a greater incentive to take risks. First, to borrow more, and, second, to invest in riskier assets that offer greater returns. Both behaviors backfire during a crisis. Lending dries up so the bank cannot replace the borrowing on which it relied, and the danger of the risky assets materializes, reducing their value and threatening the bank's solvency. One must not attribute financial crises to deposit insurance, however, because crises occurred with regularity before deposit insurance. Regulation tries to mitigate the problem by discouraging short-term borrowing and by ensuring solvency through the regulation of capital.

### 3.2.4 Capital Regulation

The capital regulation of banks seeks to make banks unlikely to fail even during a panic. Capital regulation has two sides, capital adequacy and asset quality.

Capital adequacy rules drive banks to have enough equity (misleadingly called *capital* in the language of bank regulation) to remain solvent in the face of a large drop of the bank's assets. Unlike the measures of solvency used in commerce, which focus on the ratio of debt to equity or debt to assets, bank regulation expresses capital adequacy rules as a percentage that the bank's equity (capital) must have with respect to the bank's assets. The effective rule (under both United States rules and the Basel III accord) is that banks must have equity more than 8% of their assets.

If banks were only subject to capital adequacy rules but not to asset quality rules, then the path to increased profits would be holding riskier assets because of the higher returns they would offer. The cascade of competitive pressures to take more risk during economic booms would backfire during the next recession, magnifying the crisis.

Asset quality regulation uses weighting rules for different types of assets. Riskier assets receive higher percentages and safer assets lower ones, down to zero for cash and the safest assets. Then the bank, first, calculates the amount of its assets using those weightings to derive its risk-adjusted assets, and, second, determines the amount of equity that the bank must have to comply with capital adequacy rules. The result is that a bank that holds a greater fraction and less safe assets must have more equity than a bank that holds a smaller fraction and safer assets.

An example illustrates. Compare two banks that hold only two kinds of assets, mortgage loan claims and claims on unsecured commercial loans. Mortgage claims, being secured by the mortgage, are safer and receive a 50% risk weighting. Unsecured commercial loans, being riskier, receive a 100% weighting. First Bank holds $30 of commercial loan claims and $70 of mortgage loan claims. Its risk-adjusted assets are $65. First Bank's equity must be the appropriate percentage (8%) of $65. Second Bank holds $70 of commercial loan claims and $30 of mortgage loan claims. Its risk-adjusted assets are $85. As a result of the riskier assets that Second Bank holds, Second Bank must hold more equity than the bank with the safer assets.

At the present, however, because the overall requirement for 8% equity is the same for the requirement of 8% equity on risk-adjusted assets, the consequence of the rule only applies to the different types of equity that banks can have outstanding. Banks may have 4.3% common stock, called tier 1 capital. The risk-adjustment, therefore, only limits the preferred stock (and other such securities senior to common but junior to debt) that riskier banks can have.

## 3.3     Extraordinary Measures of 2008

All these policies directly and indirectly addressing crises failed in 2008. When the brokerage house Lehman Brothers failed, not being a commercial bank, it was not subject to the FDIC's orderly liquidation. Lehman filed for bankruptcy. A run on money market mutual funds followed, because some money market funds announced that their potential exposure to Lehman prevented them from honoring withdrawal obligations at 100%. The fear, therefore, arose that the same may be the case with other money market mutual funds, if not in the case of Lehman, then in the next one. Institutional investors rushed to take their funds out of money market mutual funds that lent to businesses. Institutional investors lent the funds to the government instead.[13] The most creditworthy businesses could not borrow.[14]

The financial crisis had arisen from a path that regulation had not addressed. In reaction to the Great Depression, where the economy's lending capacity eroded because banks failed, regulation had made banks safe and protected retail depositors so that they did not have any incentive to run on their banks. The 2008 crisis, however, culminated in a run by institutional investors on money market funds, which were subject to neither bank regulation of capital, orderly liquidation, nor to deposit insurance.

When regulators saw that the financial crisis had spilled over to the real sector, they pulled all the stops. Not only did they extend the guarantee of small deposits in insured banks to all claims against all financial institutions, but they also stepped forth and directly started purchasing commercial paper, i.e., they started lending directly to businesses (rather than banks) as well as money market funds.[15] The regulators were determined not to allow the financial crisis to hamstring the borrowing of non-financial firms.

## 3.4    Post-2008 Reforms

The reforms that followed the 2008 crisis seek to limit the vulnerability of the economy to the failure of financial institutions that are not regulated banks while formalizing the reaction to a major crisis. While the reforms are voluminous (the Dodd-Frank Act that enacted them is

---

[13]Technically speaking, the money market mutual funds that lend to businesses are called prime money market mutual funds, and only those funds experienced the run. Rather than directly invest in government securities, the withdrawn funds were redirected to money market mutual funds that invest exclusively in government securities. *See* Lawrence Schmidt, Allan Timmermann & Russ Wermers, *Runs on Money Market Mutual Funds*, 106 Am. Econ. Rev. 2625 (2016).

[14]*See* Nicholas L. Georgakopoulos, The Logic of Securities Law 157–58 (2017); Simon H. Kwan, *Financial Crisis and Bank Lending* (Fed. Reserve Bank of S. F., Working Paper No. 2010-11, 2010), https://www.frbsf.org/economic-research/files/wp10-11bk.pdf.

[15]*See* Footnote 2, above.

2300 pages[16]), from the perspective of this chapter the most salient ones are the creation of the Financial Stability Oversight Council (FSOC); its power to designate nonbank financial institutions as systemically important; the orderly liquidation authority to use a procedure akin to that of the FDIC to take over a failing nonbank financial institution so as to allow it to continue to fulfill its obligations; and the Fed's power to engage in broad-based lending.[17]

The reasoning of the responses becomes visible by unfolding the events of 2008 while hypothesizing that the new regime was in place. An early casualty of the financial crisis was the Bear Stearns brokerage house, which had invested heavily in mortgage-backed securities. Rather than having been bought out by the J. P. Morgan Chase commercial bank with guarantees by the Fed about Bear's assets, Bear's failure would either have been averted if it had been designated a systemically important financial institution and been required to hold more equity, or its failure would lead to orderly liquidation, which may still have led to Bear becoming part of J.P. Morgan Chase. Then, Lehman may also have been designated a systemically important financial institution and be required to hold more equity, perhaps preventing it from failing. If Lehman failed, then it would not file for bankruptcy as it did but be subject to orderly liquidation. Hopefully, that would prevent the panic among the money market mutual funds. If the panic still arose and the economy's lending capacity declined dramatically, then the Fed would have the authority for the broad-based lending in which it engaged. Hopefully, these measures would prevent the financial crisis from causing the extensive failure of non-financial firms.[18]

---

[16]*Oversight of Dodd-Frank Act Implementation*, Fin. Serv.'s Committee, https://financialservices. house.gov/dodd-frank/ [perma.cc/BB65-JKE2] (last visited Mar. 3, 2018).

[17]*See* 12 U.S.C § 5321(a) (2012) (establishing the FSOC); 12 U.S.C. § 5322(a)(2)(J) (giving the FSOC the power to identify systemically important financial institutions); 12 U.S.C. § 5323(a) (providing authority to require supervision and regulation of nonbank financial companies in material financial distress); 12 U.S.C. § 5384 (providing necessary authority to liquidate failing financial companies); 12 U.S.C. § 343 (broad-based lending authority by the Fed in "unusual and exigent circumstances").

[18]12 U.S.C. § 343 (2012); *see also* Press Release, Fed. Reserve Bd. Approves Final Rule Specifying its Procedures for Emergency Lending Under Section 13(3) of the Fed. Reserve Act, Bd. of Governors of the Fed. Reserve Sys. (Nov. 30, 2015), https://www.federalreserve.gov/newsevents/ pressreleases/bcreg20151130a.htm.

Evaluating the response to the crisis requires one to imagine nightmare scenarios that can unfold without triggering or allowing the safety measures to work. One concern is that the FSOC, being a larger and more diverse collective body than the Fed, would not react as fast and decisively, e.g., would not admit as many institutions into the ranks of systemically important, or would not lend as broadly.[19] A different concern is that a panic may arise despite the orderly liquidations of firms like Bear or Lehman.[20]

A more dire view would assume significant changes in the economy over decades of technological developments that change dramatically the structure of the financial system (akin to the way that the development of money market mutual funds let the 2008 crisis circumvent the safety measures that were centered on the health of banks). The developments of cryptocurrencies, peer-to-peer lending, and automated contracts point to a potential such path.

Suppose that financial firms are mostly replaced by peer-to-peer lending, where savers lend directly to businesses or consumers. Suppose also that this lending occurs through cryptocurrencies, which no government can produce (although, when governments issue cryptocurrencies, that will no longer be true). Suppose, finally, that all this lending and borrowing occurs through automated contracts and algorithms, which include automated reactions to drops in the values of borrowers' assets or deterioration of borrowers' financial health. The crisis arises through a massive fraction of such clauses being triggered by some drop in the values of some class of assets. The automated reaction is to call loans (producing more insolvencies, sales, and drops of value) and cancel new loans. The result is a dramatic drop in the lending capacity of the economy in terms of cryptocurrencies and, as borrowers scramble to repay in cryptocurrencies, a dramatic rise in the value of the cryptocurrencies.

---

[19]I have taken this position previously. *See* NICHOLAS L. GEORGAKOPOULOS, THE LOGIC OF SECURITIES LAW 172 (2017); Nicholas L. Georgakopoulos, *Financial Armageddon Routs Law Again*, 14 U.C. DAVIS BUS. L. J. 1, 39 (2013).

[20]Former Treasury Secretary Tim Geithner has this concern, *see* Justin Baer & Ryan Tracy, *Ten Years After the Bear Stearns Bailout, Nobody Thinks It Would Happen Again*, WALL STREET J., https://www.wsj.com/articles/ten-years-after-the-bear-stearns-bailout-nobody-thinks-it-would-happen-again-1520959233 [perma.cc/4SZV-CJ4A] (last updated Mar. 13, 2018).

None of this chain of events touches the anti-crisis policies that exist. Having central banks supply more state-issued currencies to the market is ineffectual (and inflationary) because the obligations and the borrowing are in cryptocurrencies—the value of cryptocurrencies would merely rise in response to the additional supply of the state-issued currency. An additional difficulty that such a system would pose is that these events would unfold in computer time, in a timeframe of milliseconds. Any regulatory response could be woefully late, after clauses in automated contracts that make employment conditional on economic circumstances may have produced vast unemployment.

# 4    Conclusion

To repeat, fear crises. Our existing, complex regulation addresses the ways past crises arose. Nightmare scenarios that we can imagine should lead to new regulation addressing them.

# Closing Note

I hope you found this book instructive. If you are a researcher, I hope that you have a better understanding of ways to make graphical representations of financial ideas. If you are a student, I hope you have a better understanding of finance. I appreciate your effort and thank you for reaching this point.

N. L. Georgakopoulos, *Illustrating Finance Policy with* Mathematica, Quantitative Perspectives on Behavioral Economics and Finance, https://doi.org/10.1007/978-3-319-95372-4

# Index

Printed by Printforce, the Netherlands